PROBABILISTIC METHODS IN COMBINATORICS

1. Thomas Ferguson. Mathematical Statistics: A Decision Theoretic Approach 1967
2. Howard Tucker. A Graduate Course in Probability. 1967
3. K. R. Parthasarathy. Probability Measures on Metric Spaces. 1967
4. P. Révész. The Laws of Large Numbers. 1968
5. H. P. McKean, Jr. Stochastic Integrals. 1969
6. B. V. Gnedenko, Yu. K. Belyayev and A. D. Solovyev. Mathematical Methods of Reliability Theory. 1969
7. Demetrios A. Kappos. Probability Algebras and Stochastic Spaces. 1969, 1970
8. Ivan N. Pesin. Classical and Modern Integration Theories. 1970
9. S. Vajda. Probabilistic Programming. 1972
10. Sheldon M. Ross. Introduction to Probability Models. 1972
11. Robert B. Ash. Real Analysis and Probability. 1972
12. V. V. Fedorov. Theory of Optimal Experiments. 1972
13. K. V. Mardia. Statistics of Directional Data. 1972
14. H. Dym and H. P. McKean. Fourier Series and Integrals. 1972
15. Tatsuo Kawata. Fourier Analysis in Probability Theory. 1972
16. Fritz Oberhettinger. Fourier Transforms of Distributions and Their Inverses: A Collection of Tables. 1973
17. Paul Erdős and Joel Spencer. Probabilistic Methods in Combinatorics. 1973

In Preparation

L. E. Maistrov. Probability Theory: A Historical Sketch

PROBABILISTIC METHODS IN COMBINATORICS

by

PAUL ERDŐS

and

JOEL SPENCER

ACADEMIC PRESS
A Subsidiary of Harcourt Brace Jovanovich, Publishers
NEW YORK AND LONDON
1974

Academic Press, Inc.
111 Fifth Avenue
New York, N. Y. 10003

Distributed in the United Kingdom by:
Academic Press, Inc. (London) Ltd.
24/28 Oval Road, London NW1

Library of Congress Catalog Card No. 72–9420
For Academic Press, Inc. ISBN 0–12–240960–4

Joint edition published
by
Academic Press, New York and London
and
Akadémiai Kiadó, Budapest

Printed in Hungary

PREFACE

The field of combinatorial mathematics has undergone a period of spectacular growth in the last generation. No longer "bargain basement topology", combinatorial mathematics is now part of the standard mathematics curriculum. Research in the subject is carried out at all the major mathematical centers. The subject borrows from many fields and is applied throughout mathematics and the natural and social sciences. Combinatorial mathematics has developed solid foundations and distinct methodologies.

It is the purpose of this monograph to describe a methodology which we call the *probabilistic method*. (An initial description of the method is given in the first section.) This method has been applied to a wide variety of combinatorial problems with much success.

The probabilistic method should be a powerful tool in any combinatorial mathematician's researches. Many difficult problems may be solved almost trivially with a basic knowledge of this method. It is an unfortunate fact, however, that many researchers do not have this basic knowledge. Perhaps this is because previous uses of this method have been scattered throughout the literature. We sincerely hope that the publication of this monograph will lead to a more general use of this method.

A word about the exercises is in order. These vary greatly in difficulty though some attempt has been made to indicate especially difficult problems. Problems marked with (?) are, at present, unsolved. Solvers of such problems are urged to contact the authors. Certain problems are marked by various amounts of money [i.e. ($ 25), ($ 50)]. These problems are at the moment unsolved. The senior author has offered these amounts as prizes for the first solutions of these problems.

We should like to thank Akadémiai Kiadó and Academic Press for their publication of this monograph. The U. S. Office of Naval Research supported the junior author in the final preparation of the manuscript. We should like to thank Mrs. Bonniesue Boyd for her excellent preparation of the monograph.

Finally, the junior author wishes to thank his wife, Maryann, for her assistance, encouragement and understanding. Without her, this enterprise would have had little meaning.

Paul Erdős
Budapest, Hungary

Joel Spencer
Cambridge, Mass. USA

CONTENTS

1. TWO EXAMPLES

This book considers a *method* of proving theorems. We shall call it the probabilistic, or nonconstructive method. It is a method that is used to prove that some member of a class has a certain property without actually constructing that member. We show that, in some probability space, the probability that such a member exists is positive. The method is best described by means of examples.

We refer to Section 2 for the appropriate notation. A tournament T is transitive if $(i, j), (j, k) \in T$ implies $(i, k) \in T$. Equivalently, T is transitive iff there exists a permutation σ of the players so that $(i, j) \in T$ iff $\sigma(i) < \sigma(j)$.

Let $v(n)$ be the largest integer such that every tournament on $\{1, \ldots, n\}$ contains a transitive subtournament on $v(n)$ players. We shall show

THEOREM 1.1 (P. ERDŐS, L. MOSER [1964]): $v(n) \leq 1 + [2 \log_2 n]$.

That is, a tournament T on n "players" exists with no transitive subtournament on $v = 1 + [2 \log_2 n]$ players. (P. STEARNS (see J. W. MOON (1968) for proof) showed by induction that $v(n) \geq 1 + [\log_2 n]$.)

PROOF: Let $\mathfrak{J} = \mathfrak{J}_n$ be the class of *all* tournaments on $\{1, \ldots, n\}$, $\mathfrak{J}' =$ the class of tournaments on $\{1, \ldots, n\}$ that do contain a transitive subtournament on v players. Then

$$(1.1) \qquad \mathfrak{J}' = \bigcup_A \bigcup_\sigma \mathfrak{J}_{A,\sigma}$$

where $A \subseteq \{1, \ldots, n\}$, $|A| = v$, σ is a permutation on A, and $\mathfrak{J}_{A,\sigma}$ is the set of T such that $T|A$ is generated by σ. If $T \in \mathfrak{J}_{A,\sigma}$ the $\binom{v}{2}$ games of $T | A$ are determined. Thus

$$(1.2) \qquad |\mathfrak{J}_{A,\sigma}| = 2^{\binom{n}{2} - \binom{v}{2}}$$

and by elementary estimates

$$(1.3) \qquad |\mathfrak{J}'| \leq \sum_{A,\sigma} 2^{\binom{n}{2} - \binom{v}{2}} = \binom{n}{v} v!\, 2^{\binom{n}{2} - \binom{v}{2}} < 2^{\binom{n}{2}} = |\mathfrak{J}|.$$

Thus $\mathfrak{J} - \mathfrak{J}' \neq \emptyset$. That is, there exists $T \in \mathfrak{J} - \mathfrak{J}'$ not containing a transitive subtournament on v players. Q.E.D.

The last line of the proof emphasizes its nonconstructive character. We prove that a set is nonempty and therefore necessarily contains the desired member T. The proof contains no algorithm for "constructing" this T. Once a theorem has been proven by nonconstructive methods, one is still interested in a proof by construction. In this example it would be of interest to "construct" a T on $\{1, \ldots, n\}$ that contained no transitive subtournament on 100 log n players, although the above proof implies that "almost all" T have this property.

A similarity exists between the nonconstructive method and the use of the Axiom of Choice. One might even think of the nonconstructive method as a finitistic analogue of the Axiom of Choice. However, there is one crucial distinction. We *can* always find the guaranteed T by laboriously checking each member of (the finite set) $\mathfrak{T}$ until we find one with the desired property. Consequently, unlike the Axiom of Choice, there are no logical difficulties.

Application of the probabilistic method to "real world" problems may pose further difficulties. For example, a computer scientist is told to construct, on a computer, the following: the matrix of a tournament on 127 players that does not contain a transitive subtournament on 15 players. Imagine that if his matrix does contain a transitive subtournament on 15 players his company (or country, or mutual fund) will suffer grave financial consequences. He learns from this book that such a tournament exists. But this is not sufficient—his boss wants a specific matrix stored in the computer. Our scientist calculates that it may cost millions of dollars (or tens of millions of forints) to find such a T by exhaustive methods. However, like many of us, he is on a limited budget. What is he going to do? If his moral character is sufficiently low he would construct T at random, pack his bags to prepare for a hasty departure, and hope for the best. Given our experiences with computer scientists, we surmise that this would be the "real world" solution. If, on the other hand, he is of outstanding moral character his only recourse is to ask for a refund on this book. We would, of course, refuse.

The "real world" situation is different if it is required that the tournament does not contain a transitive subtournament on $v = 16$ players. Then

$$\frac{|\mathfrak{T}'|}{|\mathfrak{T}|} \leq \binom{127}{16} 2^{-\binom{16}{2}} \sim .00128\,!$$

A "random" tournament will have the desired property with probability $\geq .998$. This probability of success is sufficient in many practical situation—our computer scientist can randomize and relax.

We shall state our theorems in the form "there exists a tournament (or graph, or matrix) with a given property." In many cases they could be stated "a 'random' tournament has a given property with probability p" where $1 - p \ll 1$. Statements of this second type might have greater applicability to other problems.

It may be argued whether or not a nonconstructive proof really gives a solution to a given problem. As indicated by our example it depends on what the real problem is.

We will now reprove Theorem 1.1 in a different light. Let $\mathbf{T} = \mathbf{T}_n$ be a random variable. Its values are the members of $\mathfrak{I}$ where for each $T \in \mathfrak{I}$, Prob $(\mathbf{T} = T) = 2^{-\binom{n}{2}}$. That is, all members of $\mathfrak{I}$ are equally probable values of $\mathbf{T}$. We can view $\mathbf{T}$ in a somewhat different way. For each pair i, j of players Prob $((i, j) \in \mathbf{T}) = 1/2$. These probabilities, taken over each game, are mutually independent. The above properties determine $\mathbf{T}$. Then

Prob ($\mathbf{T}$ contains a transitive subtournament on v players)

$$\leq \sum_A \sum_\sigma \text{Prob}\ (\mathbf{T} \mid A \text{ generated by } \sigma) \tag{1.4}$$

$$= \binom{n}{v} v!\, 2^{-\binom{v}{2}} < 1\,.$$

Thus some value T of $\mathbf{T}$ does not contain a transitive subtournament on v players.

We have basically rewritten the original proof. The original proof is an example of "counting argument". The second proof is a "random variable argument". Most theorems in this book could be written in either style. Random variable arguments seem clearer and more intuitive. In more complicated results (e. g., Theorems 5.3, 8.1) the random variable style allows direct use of probability theorems. We shall tend toward the use of random variables. We note, however, that most previous work has used counting arguments.

Consider one more example. A Hamiltonian path P in a tournament T_n is defined as a permutation $P(1), \ldots, P(n)$ of the players in the tournament such that for $1 \leq i < n$, $(P(i), P(i+1)) \in T$. The following result, due to T. Szele (1943), was perhaps the first use of nonconstructive methods.

Theorem 1.2: *There exists* $T = T_n$ *that contains* $\geq n!\, 2^{-n+1}$ *Hamiltonian paths.*

Proof: Let $\mathfrak{I} = \mathfrak{I}_n$ be as before. Set

$$U = \{(T, P) : T \in \mathfrak{I}_n, P \text{ is a Hamiltonian path in } T\}$$

$$|U| = \sum_T |\{P : (T, P) \in U\}| \tag{1.5}$$

$$= \sum_P |\{T : (T, P) \in U\}|\,. \tag{1.6}$$

Fix any path P. Then $(T, P) \in U$ iff a certain $(n-1)$ games of T are determined. Thus, $(T, P) \in U$ for exactly $2^{\binom{n}{2} - (n-1)}$ tournaments T,

Since there are $n!$ paths P,

$$|U| = n!\, 2^{\binom{n}{2} - (n-1)}. \tag{1.7}$$

The right-hand side of (1.5) has exactly $2^{\binom{n}{2}}$ addends. Therefore, for some T,

$$|\{P : (T, P) \in U\}| \geq n!\, 2^{-(n-1)}. \quad \text{Q.E.D.} \tag{1.8}$$

(Note: T. SZELE also showed, by constructive methods, that the maximum number of Hamiltonian paths is bounded from above by $n!\, 2^{-3n/4}$.)

Now we rewrite this proof as a random variable argument. Again, let $\mathbf{T} = \mathbf{T}_n$ be a random member of $\mathfrak{J}_n$, having the same distribution as before. Define f_P on $\mathfrak{J}_n$ by

$$f_P(T) = \begin{cases} 1 \text{ if } P \text{ is a Hamiltonian path in } T \\ 0 \text{ otherwise.} \end{cases}$$

Set

$$N(T) = \sum_P f_P(T) = \text{the number of Hamiltonian paths in } T.$$

Then

$$\mathsf{E}[N(\mathbf{T})] = \mathsf{E}[\sum_P f_P(\mathbf{T})] \tag{1.9}$$

$$= \sum_P \mathsf{E}[f_P(\mathbf{T})] \tag{1.10}$$

$$= \sum_P 2^{-n+1} \tag{1.11}$$

$$= n!\, 2^{-n+1} \tag{1.12}$$

where E represents the expected value. Thus, some $N(T) \geq n!\, 2^{-n+1}$.

The crucial step from (1.9) to (1.10) requires

THEOREM 1.3: *If* $\mathbf{X}_1, \ldots, \mathbf{X}_n$ *are random variables with expectations, then the expectation of their sum exists and is the sum of their expectations.*

A proof may be found in any elementary probability text (e. g., W. FELLER [1957] p. 208). Notice that no assumptions are made on the dependence or independence of the $\mathbf{X}_i$. The surprising power of Theorem 1.3 is indicated by Exercises 3 and 4.

This use of expected value is called the first-moment method.

Exercises

1. Let $w(n)$ be the largest integer so that for every tournament T on $\{1, \ldots, n\}$ there exist disjoint sets $A, B \subseteq \{1, \ldots, n\}$ of cardinality $w(n)$ such that $A \times B \subseteq T$. Find an upper bound for $w(n)$.

2. A Hamiltonian cycle in a tournament E is a permutation P such that $(P(i),\ P(i+1)) \in E$ for $1 \leq i < n$ and $(P(n),\ P(1)) \in T$. Prove that there exists a tournament T that contains at least $n!\,2^{-n}$ Hamiltonian cycles.

3. A hat-check girl at a night club is given 30 hats. She then has a dispute with the manager and so, to spite him, returns the 30 hats at random. What is the expected number of people who get their own hats returned?

4. The nation of Northumbria decides to give special license plates, with numbers 1 to 2000 inclusive, to the nation's political, business, and spiritual leaders. After much dispute the list of 2000 leaders is determined. It is decided, to avoid further arguments, to distribute the 2000 plates at random to the leaders. What is the expected number of leaders whose license plate number equals the year of their birth?

2. NOTATION

In this section, we give the notation that shall be standard throughout the book. This section is intended mainly for reference, though the reader is advised to skim through it at first reading. The probabilistic notation is somewhat unusual and essential to the book.

Throughout the book (except §17A) we deal exclusively with finite objects.

There is some ambiguity in the notation (e. g., $[x]$) but the usage shall (hopefully) be clear by the context.

Arithmetic

If x is real

$$[x] = \text{largest integer} \leq x$$

$$\{x\} = \text{smallest integer} \geq x.$$

If $t \geq 1$, integral

$$t\,! = \prod_{i=1}^{t} i$$

$$0\,! = 1$$

$$\binom{x}{t} = 0 \text{ if } x \leq t-1$$

$$= \left[\prod_{i=0}^{t-1} (x-i)\right] \Big/ t\,! \text{ if } x \geq t-1$$

$$\binom{x}{0} = 1 \text{ for all } x.$$

Sets

If X is a set

$$|X| = \text{cardinality } (X) = \text{number of elements of } X.$$

$$2^X = \text{power set } (X) = \{Y\colon Y \subseteq X\}.$$

$$[X]^r = \{Y : Y \subseteq X, |Y| = r\}$$

$[n] = \{x : x \text{ integral}, \ 1 \leq x \leq n\} = \{1, \ldots, n\}$, often used as a "standard" n element set.

$[n]^r = [[n]]^r = \{Y : Y \subseteq [n], \mid Y \mid = r\}$

$\cap$ = set intersection

$\cup$ = set union

$X - Y = \{x : x \in X, x \notin Y\}$

$X \Delta Y = (X \cup Y) - (X \cap Y)$ = symmetric difference of X, Y.

iff = if and only if $(X \Delta Y = \{x : x \in X \text{ iff } x \in Y\})$.

Graphs

A graph G on a set of vertices V is a subset of $[V]^2$. The elements of G are called edges. If $x \in G$, degree $(x) = \mid \{y \in V : \{x, y\} \in G\} \mid$. If $W \subseteq V$ then $G \mid W = G \cap [W]^2$ is called the subgraph generated by W. A subgraph H of G is a subset of G. A set $W \subseteq V$ is called independent if $G \cap [W]^2 = \emptyset$. W is called a clique if $G \cap [W]^2 = [W]^2$. The clique number of G, denoted by $\omega(G)$, is the maximal $\mid W \mid$ where W is a clique of G.

Coloration

An r-coloration of a graph G is a map $h : G \to [r]$ such that for all $e = \{x, y\} \in G$, $h(x) \neq h(y)$. The minimal r for which an r-coloration of G exists is called the chromatic number of G and denoted by $\chi(G)$. It shall sometimes be convenient to consider 2-colorations as maps $h : G \to \{+1, -1\}$.

Hypergraphs

A hypergraph G on a set of vertices V is a subset of 2^V. The elements of G are called hyperedges. A k-graph G is a subset of $[V]^k$. So 2-graphs are graphs. If G is a k-graph, $W \subseteq V$ and $G \mid W = [W]^k$ W is called a clique. The clique number $\omega(G)$ of a k-graph is the maximal $\mid W \mid$ where W is a clique. If G is a hypergraph an r coloration of G is a map $h : G \to [r]$ so that for all $e \in G$ there exist $x, y \in e$, $h(x) \neq h(y)$. The chromatic number $\chi(G)$ is defined as for graphs, as are sets of independent points.

Tournaments

A digraph D on a set V is a set of ordered pairs (x, y), $x, y \in V$, $x \neq y$. A tournament T on a set V is a digraph such that for all x, $y \in V$, $x \neq y$ either $(x, y) \in T$ or $(y, x) \in T$ but not both. Intuitively, V is thought of as a set of players and $(x, y) \in T$ denotes x having defeated y. The elements of $[V]^2$ are called the games of T.

If a graph, hypergraph, digraph, or tournament is on the set $V = [n]$ we may denote this by G_n, K_n, T_n.

Probability

Now we want to adjoin the notational structure of probability to that already given. We shall denote random variables by using *boldface* (e. g., $\mathbf{x}$, $\mathbf{X}$, $\mathbf{G}$, $\mathbf{T}$). As we deal with finite objects our random variables take on only a finite number of possibilities. That is, given $\mathbf{x}$, there exist $x_1, \ldots, x_s$, $0 \leq p_1, \ldots, p_s \leq 1 = \Sigma p_i$, Prob $[\mathbf{x} = x_i] = p_i$.

Random Graphs

For $0 \leq p \leq 1$, $\mathbf{G}_{n,p}$ is a random variable whose values are graphs on $[n]$. If $e \in [n]^2$, Prob $[e \in \mathbf{G}_{n,p}] = p$ and these probabilities are independent for different e. Thus

$$\text{Prob}\,[\mathbf{G}_{n,p} = G] = p^{|G|}(1-p)^{\binom{n}{2} - |G|}.$$

For $0 \leq e \leq \binom{n}{2}$, $\mathbf{G}_{n,e}$ is a random member of $[[n]^2]^e$. That is, $\mathbf{G}_{n,e}$ is a random graph with n points, e edges. $\mathbf{G}_{n,p}$ and $\mathbf{G}_{n,e}$ will be distinguished by the letter used.

Random Tournaments

$\mathbf{T}_n$ is a random tournament on $[n]$. For all T_n, Prob $[\mathbf{T}_n = T_n] = 2^{-\binom{n}{2}}$. If $x, y \in [1, n]$, $x \neq y$, Prob $[(x, y) \in \mathbf{T}_n] = 1/2$. These probabilities are independent except $(x, y) \in \mathbf{T}_n$ iff $(y, x) \notin \mathbf{T}_n$.

Random Objects

For convenience we speak of $\mathbf{x}$ as a random member of a class C if each $x \in C$ is achieved with equal probability. For example, "$\mathbf{S}$ is a random subset of $[n]$ of cardinality r" means that for $S \subseteq [n]$, $|S| = r$,

$$\text{Prob}\,[\mathbf{S} = S] = \binom{n}{r}^{-1}.$$

o, 0

We shall sometimes use the o, 0 notation. If $f(n)$, $g(n)$ are two functions we write $f(n) = o(g(n))$ if $\lim\limits_{n \to \infty} f(n)/g(n) = 0$. We write $f(n) = 0(g(n))$ if there exists C, n_0, so that for $n > n_0$ $|f(n)/g(n)| < C$. We shall generally use n for the number of vertices in our collection. If the o, 0 notation is used without a variable, that variable is assumed to be n.

3. BINOMIAL DISTRIBUTION

In this section we state some results on the binomial distribution. These results are used (usually unreferenced) throughout the book.

Let $\mathbf{X}_i$, $1 \leq i \leq n$, be independent random variables with Prob $(\mathbf{X}_i = +1) = \text{Prob}\,(\mathbf{X}_i = -1) = 1/2$. We define

$$\mathbf{S}_n = \sum_{i=1}^{n} \mathbf{X}_i \,. \tag{3.1}$$

Let $\mathbf{Y}_i$, $1 \leq i \leq n$, be independent random variables with Prob $(\mathbf{Y}_i = 1) = \text{Prob}\,(\mathbf{Y}_i = 0) = 1/2$. We define

$$\mathbf{U}_n = \sum_{i=1}^{n} \mathbf{Y}_i \,. \tag{3.2}$$

Then

$$\text{Prob}\,[\mathbf{U}_n = i] = \binom{n}{i} p^i (1-p)^{n-i}\,, \quad 1 \leq i \leq n\,. \tag{3.3}$$

We require the well-known Stirling's Formula

$$n\,! = n^n e^{-n} \sqrt{2\pi n}\,\big(1 + o(1)\big)\,. \tag{3.4}$$

H. ROBBINS [1955] refined (3.4) to

$$n^n e^{-n} \sqrt{2\pi n}\, e^{1/(12n+1)} < n\,! < n^n e^{-n} \sqrt{2\pi n}\, e^{1/12n}\,. \tag{3.5}$$

Though we shall never make use of (3.5) in this book, it is occasionally useful for proving a theorem (e. g., Theorem 15.1) for all n.

If $\mathbf{X}$ and $\mathbf{Y}$ are two random variables we write $\mathbf{X} \sim \mathbf{Y}$ if they have the same distributions. Thus $2\mathbf{U}_n - n \sim \mathbf{S}_n$.

We shall use the common notation $B(n, p)$ for the binomial distribution. That is, $B(n, p)$ is the distribution of the number of successes in n independent trials, with probability of success p. So $\mathbf{U}_n$ has distribution $B(n,\ 1/2)$.

Let $N(\mu, \sigma)$ denote the normal distribution of mean μ and standard distribution σ. For large n, $\mathbf{S}_n$ can be approximated by $N(0, \sqrt{n})$. In particular

(3.6) If $\lim_n \alpha(n)\, n^{-1/2} = c$ then

$$\lim_n \text{Prob}\,[\mathbf{S}_n \geq \alpha(n)] = \Phi(c) = \frac{1}{\sqrt{2\pi}} \int_c^{\infty} e^{-x^2/2}\, dx\,.$$

We are particularly interested in the tail of the binomial distribution. Let p, q be positive integers such that $p + q = 1$. H. CHERNOFF [1952] has shown

$$(3.7)\quad \sum \binom{m}{t} p^t q^{m-t} \leq \exp[(m-k)\log(mq/(m-k)) + k\log(mp/k)]$$

where the sum either is over t such that $t \geq k$ or over t such that $t \leq k$ provided that $k \geq pm$ or $k \leq pm$. J. W. MOON [1971] sets $p = q = \frac{1}{2}$, $k = \frac{1}{2} m + \lambda$ to show

$$(3.8)\qquad \sum_{t > \frac{1}{2}m + \lambda} \binom{m}{t} < 2^m \exp[-2\lambda^2/m]$$

given $0 \leq \lambda \leq \frac{1}{2} m$. In our notation, if $\lambda > 0$

$$(3.9)\qquad \text{Prob}\,[\mathbf{S}_m > \lambda] < e^{-2\lambda^2/m}\,.$$

This formula is used throughout the book.

We shall often also use the following elementary asymptotic formula

$$(3.10)\qquad \binom{n}{k} \sim \frac{n^k e^{-\frac{k^2}{2n} - \frac{k^3}{6n^2}}}{k!}(1 + o(1)) \text{ valid for } k = o(n^{3/4})\,.$$

4. THE PROPERTY B

Let $\mathfrak{I} = \{A_a\}$ be a hypergraph, $M = \cup A_a$.

Definition: $\mathfrak{I}$ is said to have *property B* if $\chi(\mathfrak{I}) \leq 2$. Equivalently, $\mathfrak{I}$ has property B if there exists $S \subseteq M$, $\emptyset \subsetneqq S \cap A \subsetneqq A$ for all $A \in \mathfrak{I}$.

E. W. Miller [1937] labelled this property. He used the letter B in honor of Felix Bernstein. P. Erdős and A. Hajnal [1961] define $m(n)$ as the minimal $|\mathfrak{I}|$ of an n-graph $\mathfrak{I}$ that does not have property B. Since $\mathfrak{I} = [2n-1]^n$ is such an n-graph, $m(n)$ exists.

Theorem 4.1: $m(n) \geq 2^{n-1}$.

Proof: Let $\mathfrak{I}$ be fixed, $|\mathfrak{I}| = f$. Let $C = (K_1, K_2)$ be a 2-coloring of M. That is, $K_1 \cup K_2 = M$, $K_1 \cap K_2 = \emptyset$. Let $\beta(C)$ be the number of monochromatic $A \in \mathfrak{I}$ under C.

$$\beta(C) = \sum_{A \in \mathfrak{I}} \sum_{i=1}^{2} \sum_{A \subseteq K_i} 1 . \tag{4.1}$$

Now let $\mathbf{C} = (\mathbf{K}_1, \mathbf{K}_2)$ be a random 2-coloring. Then

$$\begin{aligned} \mathsf{E}[\beta(\mathbf{C})] &= \sum_{A \in \mathfrak{I}} \sum_{i=1}^{2} \text{Prob}\,[A \subseteq \mathbf{K}_i] \\ &= \sum_{A \in \mathfrak{I}} \sum_{i=1}^{2} 2^{-n} \quad (\text{as } |A| = n) \qquad (4.2) \\ &= f 2^{-n+1} . \end{aligned}$$

If $\mathfrak{I}$ does not have property B then for all C, $\beta(C) \geq 1$. Thus

$$1 \leq \mathsf{E}[\beta(\mathbf{C})] = f 2^{-n+1} \tag{4.3}$$

and therefore

$$2^{n-1} \leq f . \tag{4.4}$$

We improve this result with a method due to W. M. Schmidt. Let $\mathfrak{I}$ be an n-graph without property B. In Theorem 4.1 we attempted a random coloring of M. If, in the random coloration of M, some $N \in \mathfrak{I}$ is monochromatic, we shall shift the color of some $x_N \in N$. We assign to each $N \in \mathfrak{I}$ an $x_N \in N$ so that

$$|\{L \in \mathfrak{I},\ L \cap N = \{x_N\}\}| < f n^{-1} . \tag{4.5}$$

Some $x \in N$ satisfies (4.5) as otherwise

$$|\mathfrak{J}| > |\{L \in \mathfrak{J} : |L \cap N| = 1\}| \geq n(fn^{-1}).$$

Given any coloring $C = (K_1, K_2)$ either $\beta(C) \geq 2$ or there exists a unique $N \in \mathfrak{J}$, $i = 1$ or 2, $N \subseteq K_i$. Set $C' = (K_1 \Delta \{x_N\}, K_2 \Delta \{x_N\})$. As $\beta(C') \geq 1$ some $L \in \mathfrak{J}$ is monochromatic in C'. This can only occur if $L \cap K_i = \{x_N\}$, which implies $L \cap N = \{x_N\}$. Thus, for all $C = (K_1, K_2)$

$$\frac{\beta(C)}{2} + \sum_{N \in \mathfrak{J}} \sum_{\substack{L \in \mathfrak{J} \\ L \cap N = \{x_N\}}} \sum_{i=1}^{2} \sum_{\substack{N \subseteq K_i \\ L \cap K_i = \{x_N\}}} \frac{1}{2} \geq 1. \tag{4.6}$$

Let $\mathbf{C}$ be a random coloring. $\mathsf{E}[\beta(\mathbf{C})] = f2^{-n+1}$ as before. There are less than $f(fn^{-1})$ (2) choices for N, L, i. For each choice

$$\text{Prob}\,[N \subseteq \mathbf{K}_i, L \cap \mathbf{K}_i = \{x_N\}] = 2^{-(2n-1)}. \tag{4.7}$$

Thus

$$\mathsf{E}[\sum_{(1)} \sum_{(2)} 1/2] < f(fn^{-1})\,(2)\,2^{-(2n-1)}\,2^{-1} \tag{4.8}$$

where $\sum_{(1)}$ is over $N, L \in \mathfrak{J}$, $N \cap L = \{x_N\}$, $i = 1$ or 2 and $\sum_{(2)}$ is over $N \subseteq \mathbf{K}_i$, $L \cap \mathbf{K}_i = \{x_N\}$. Taking expected values in (4.6),

$$f2^{-n} + f^2 n^{-1} 2^{-(2n-1)} \geq 1 \tag{4.9}$$

which implies

THEOREM 4.2 (W. M. SCHMIDT [1964]): $m(n) \geq 2^n(1 + 2n^{-1})^{-1}$.

THEOREM 4.3 (P. ERDŐS [1964]): $m(n) \leq (1 + o(1))\, e\, (\log 2)\, n^2 2^{n-2}$.

PROOF: Set $r = [n^2/2]$. Let $\mathbf{A}_1, \mathbf{A}_2, \ldots$ be independent random members of $[r]^n$. Set $\mathbf{F} = \mathbf{F}_f = \{\mathbf{A}_1, \ldots, \mathbf{A}_f\}$. For any coloration $C = (K_1, K_2)$, and any i

$$\text{Prob}\,(\mathbf{A}_i \text{ is monochromatic}) = \left[\binom{|K_1|}{n} + \binom{|K_2|}{n}\right] \Big/ \binom{r}{n}$$

$$> 2\binom{r/2}{n} \Big/ \binom{r}{n} \quad \text{[by Lemma 12.2]} \tag{4.10}$$

$$> 2e^{-1}\, 2^{-n}(1 + o(1))$$

by elementary calculations. Since the $\mathbf{A}_i$ are independent

$$\text{Prob}\,(\mathbf{F} \text{ is colored by } C) < \left(1 - 2e^{-1}\, 2^{-n}(1 + o(1))\right)^f \tag{4.11}$$

and therefore

$$\mathsf{E}[|\{C : \mathbf{F} \text{ is 2-colored by } C\}|] < 2^r\left(1 - 2e^{-1}\, 2^{-n}(1 + o(1))\right)^f. \tag{4.12}$$

Set $f = (1 + o(1))\, e (\log 2)\, n^2 2^{n-2}$. The expected number of 2-colorations of **F** is less than unity. Therefore there exists F_f, $| F_f | \leq f$, that cannot be 2-colored.

Exercises

P. Erdős [1969] defines $m_N(n)$ $(N \geq 2n - 1)$ as the minimal $| \mathfrak{I} |$ of an n-graph $\mathfrak{I}$ not having property B where $N = | \cup \mathfrak{I} |$.

1. Use the method of Theorem 4.2 to show

$$m_{2N}(n) \leq -2N \log 2/\log\left[1 - 2\binom{N}{n}\Big/\binom{2N}{n}\right].$$

2. Show

$$m_{2N}(n) \geq \binom{2N}{n}\Big/2\binom{N}{n}.$$

(Hint: Let $\mathbf{C} = (\mathbf{K}_1, \mathbf{K}_2)$ be a 2-coloring of $\cup \mathfrak{I}$, chosen equiprobably among those colorations with $| K_1 | = | K_2 | = N$. Use the method of Theorem 4.1.)

3. Does there exist an n-graph $\mathfrak{I}$ which does not have property B such that $A_1, A_2 \in \mathfrak{I}$ imply $| A_1 \cap A_2 | \leq 1$? (This problem may be very difficult without the methods of Theorem 11.2. See § 11, Exercise 4, for generalization. See H. Abbott [1965] for a constructive proof.)

4. (?) Given k, do there exist $n < m$ and a family $F \subseteq [m]^n$ with the following properties:

(a) F does not have property B.
(b) F is a minimal family not having property B.
That is, if $A \in F$, $F - \{A\}$ has property B.
(c) $A, B \in F$ implies $| A \cap B | \leq 1$.
(d) For all $x \in [m]$

$$| \{A : x \in A \in F\} | \geq k.$$

This question is due to B. Toft.

5. M. Herzog and J. Schönheim [1972] define $\mathfrak{F}$ to have property B_r if $\chi(\mathfrak{F}) \leq r$. They set $m_r(p) =$ the minimal $| \mathfrak{F} |$ such that $\mathfrak{F}$ is a p-graph not having property B_r. Find bounds analogous to Theorems 4.2, 4.3 for $m_r(p)$, r fixed. In particular, prove

$$\lim_{p \to \infty} m_r(p)^{1/p} = r$$

6. Tighten (4.9) to

$$f 2^{-n} + f[(f - 1)/n]\, 2^{-(2n-1)} \geq 1$$

and use this to show $m(3) > 6$.

5. RAMSEY'S THEOREM

The following theorem, first stated and proven by F. P. RAMSEY [1930], is the basis for many combinatorial investigations.

RAMSEY'S THEOREM (unabridged): *Given integers* $s, k_1, \ldots, k_j$ *there exists a minimal integer* $r = R_s(k_1, \ldots, k_j)$ *with the following property:*

If $|S| \geq r$ *and* $C = (K_1, \ldots, K_j)$ *is a* j*-coloring of* $[S]^s$ *then for some* i, $1 \leq i \leq j$, *there exists* A, $|A| = k_i$, $[A]^s \subseteq K_i$.

Proofs of RAMSEY's Theorem in various forms may be found in many combinatorial texts. (See H. J. RYSER [1963] for the above theorem.) We shall mainly be interested in the special case $s = j = 2$.

RAMSEY'S THEOREM (abridged): *Given integers* $k, l \geq 2$ *there exists a minimal integer* $r = R(k, l)$ *with the following property:*

If $|S| = r$, *and* $C = (K_1, K_2)$ *is a 2-coloring of* $[S]^2$ *then either*

(a) *there exists* $T \subseteq S$, $|T| = k$, $[T]^2 \subseteq K_1$

or

(b) *there exists* $U \subseteq S$, $|U| = l$, $[U]^2 \subseteq K_2$.

PROOF: Clearly $R(k, 2) = R(2, k) = k$. Let $k, l > 2$ and assume, by induction, that $R(k-1, l)$ and $R(k, l-1)$ exist. Set $r = R(k-1, l) + R(k, l-1)$ and let $|S| \geq r$ and $C = (K_1, K_2)$ be a coloring of $[S]^2$. Fix $x \in S$ and set

$$A = \{y \in S : \{x, y\} \in K_1\},$$

$$B = \{y \in S : \{x, y\} \in K_2\}.$$

As $|A| + |B| = r - 1$ either $|A| \geq R(k-1, l)$ or $|B| \geq R(k, l-1)$. Assume the former (the latter case is symmetric).

By induction either

(b′) there exists $U \subseteq A$, $|U| = l$, $[U]^2 \subseteq K_2$. But then $U \subseteq S$ so (b) is satisfied

or

(a′) there exists $T_1 \subseteq A$, $|T_1| = k - 1$, $[T_1]^2 \subseteq K_1$.
But then set $T = T_1 \cup \{x\}$.
$|T| = k$ and $[T]^2 \subseteq K_1$ so (a) satisfied. Q.E.D.

Since

$$R(k, l) \leq R(k-1, l) + R(k, l-1) \tag{5.1}$$

we can show by induction

$$R(k, l) \leq \binom{k+l-2}{k-1}. \tag{5.2}$$

We first consider lower bounds on $R(k, l)$. Suppose $k = l$ and $r = R(k, k)$. Set $M = [r]^2$ and

$$\mathfrak{J} = \{[T]^2 : |T| = k, T \subseteq [r]\}. \tag{5.3}$$

Then $\mathfrak{J}$ does not have property B. Since $\mathfrak{J}$ is a $\binom{k}{2}$-graph, by Theorem 4.1

$$\binom{r}{k} = |\mathfrak{J}| \geq m\left(\binom{k}{2}\right) \geq 2^{\binom{k}{2}-1}. \tag{5.4}$$

Applying elementary estimates to (5.4) yields

THEOREM 5.1 (P. ERDŐS [1947]): $R(k, k) \geq k2^{k/2}\left[\frac{1}{e\sqrt{2}} - o(1)\right]$.

Now fix k, l. Let n be an integer and $\mathbf{C} = (\mathbf{K}_1, \mathbf{K}_2)$ be a 2-coloring of $[n]^2$ such that $\mathbf{K}_1 = \mathbf{G}_{n,p}$ where $0 \leq p \leq 1$. If $A \in [n]^k$ then Prob $[[A]^2 \subseteq \mathbf{K}_1] = p^{\binom{k}{2}}$. If $A \in [n]^l$ then Prob $[[A]^2 \subseteq \mathbf{K}_2] = (1-p)^{\binom{l}{2}}$ since $\mathbf{K}_2$ has distribution $\mathbf{G}_{n,1-p}$. Set $\beta(C) =$ the number of k-sets whose edges are all "1" plus the number of l-sets whose edges are all "2". Then

$$\mathsf{E}[\beta(\mathbf{C})] = \binom{n}{k} p^{\binom{k}{2}} + \binom{n}{l} (1-p)^{\binom{l}{2}}. \tag{5.5}$$

If $\mathsf{E}[\beta(\mathbf{C})] < 1$ there exists C, $\beta(C) = 0$ so $n \leq R(k, l)$. This proves

THEOREM 5.2: *If there exists p, $0 \leq p \leq 1$ such that*

$$\binom{n}{k} p^{\binom{k}{2}} + \binom{n}{l} (1-p)^{\binom{l}{2}} < 1$$

then $n \leq R(k, l)$.

Now fix $k = 3$. By (5.2), $R(3, l) \leq \binom{l+1}{2} \sim l^2/2$.

THEOREM 5.3 (P. ERDŐS [1961]): $R(3, l) \geq c_1 l^2/(\log l)^2$.

PROOF: Due to the monotonicity of R it is sufficient to show

$$R(3, c_2\sqrt{n}\log n) \geq n. \tag{5.6}$$

Consider a coloring $C = (K_1, K_2)$ as a graph $G = K_1$. We need to find G on $[n]$ containing no triangle and no set of $x = [c_2 \sqrt{n} \log n]$ independent points. First, we weaken this condition. If $X \in [n]^x$ write $\alpha(X, G)$ if $\exists i, j \in X$, $\{i, j\} \in G$ so that $k \notin X$ implies $\{i, k\} \notin G$ or $\{j, k\} \notin G$. Intuitively, $\alpha(X, G)$ means there exists an edge in X not extendable to a triangle outside of X.

LEMMA 5.1: *If $\alpha(X, G)$ for all $X \in [n]^x$ there exists $H \subseteq G$ containing no triangles and no set of x independent points.*

PROOF: Say $G = \{e_1, \ldots, e_s\}$. Set $H_0 = \emptyset$. For $1 \leq i \leq s$ set $H_i = H_{i-1} \cup \{e_i\}$ if $H_{i-1} \cup \{e_i\}$ contains no triangles, otherwise set $H_i = H_{i-1}$. Set $H = H_s$. By definition H contains no triangles and $H \subseteq G$. Now let $X \in [n]^x$. Because $\alpha(X, G)$ we find $e_t = \{i, j\} \in G$ not extendable to a triangle outside of X. So if $e_t \notin H$, $H_{t-1} \cup \{e_t\}$ contains a triangle, so for some k, $\{i, k\}$ and $\{j, k\} \in H_{t-1} \subseteq G$. As $\alpha(X, G)$ we must have $k \in X$ so $\{i, k\} \in H \cap [X]^2$ and X is not independent.

To prove Theorem 5.3 we need only show

LEMMA 5.2.: *Let $\mathbf{G} = \mathbf{G}_{n,p}$ where $p = 2c_2^{-1/2}n^{-1/2}$. For c_2 sufficiently large*

$$\text{Prob}\left[\alpha(X, \mathbf{G}) \text{ for all } X \in [n]^x\right] > 0 .$$

PROOF: Fix $X \in [n]^x$. It suffices to show

$$\text{Prob}\,[\text{not } \alpha(X, \mathbf{G})] < \binom{n}{x}^{-1}. \tag{5.7}$$

Set

$$\mathbf{H}_1 = \bigcup_{i \notin X} \{j \in X : \{i, j\} \in \mathbf{G}\}^2$$
$$\mathbf{H}_2 = \mathbf{G} \mid X. \tag{5.8}$$

Then (5.7) is equivalent to

$$\text{Prob}\,[\mathbf{H}_2 \subseteq \mathbf{H}_1] < \binom{n}{x}^{-1}. \tag{5.9}$$

By Adam's Theorem*

* Adam's Theorem states that if events B , $1 \leq i \leq t$, are mutually exclusive and mutually exhaustive

$$\text{Prob}\ (A) = \sum_{i=1}^{t} \text{Prob}\ (A \mid B_i)\ \text{Prob}\ (B_i) .$$

As a corollary,

$$\text{Prob}\ (A) \leq \max_i \text{Prob}\ (A \mid B_i) .$$

The result is called Adam's Theorem because it has been rediscovered so often.

$$\text{Prob}\,[\mathbf{H}_2 \subseteq \mathbf{H}_1] = \text{Prob}\left[\mathbf{H}_2 \subseteq \mathbf{H}_1 \,\middle|\, |\mathbf{H}_1| \leq \frac{1}{2}\binom{x}{2}\right]$$

$$\text{Prob}\left[|\mathbf{H}_1| \leq \frac{1}{2}\binom{x}{2}\right] +$$

$$\text{Prob}\left[\mathbf{H}_2 \subseteq \mathbf{H}_1 \,\middle|\, |\mathbf{H}_1| > \frac{1}{2}\binom{x}{2}\right]$$

(5.10)

$$\text{Prob}\left[|\mathbf{H}_1| > \frac{1}{2}\binom{x}{2}\right] <$$

$$\text{Prob}\left[\mathbf{H}_2 \subseteq \mathbf{H}_1 \,\middle|\, |\mathbf{H}_1| \leq \frac{1}{2}\binom{x}{2}\right] +$$

$$\text{Prob}\left[|\mathbf{H}_1| > \frac{1}{2}\binom{x}{2}\right].$$

To show (5.9) we bound the addends of (5.10). By Adam's Theorem

$$\text{Prob}\left[\mathbf{H}_2 \subseteq \mathbf{H}_1 \,\middle|\, |\mathbf{H}_1| \leq \frac{1}{2}\binom{x}{2}\right] \leq \max \text{Prob}\left[\mathbf{H}_2 \subseteq \mathbf{H}_1 \,\middle|\, \mathbf{H}_1 = H_1\right]$$

(5.11)

$$\left(\max \text{ over } H_1 \subseteq [x]^2,\ |H_1| \leq \frac{1}{2}\binom{x}{2}\right)$$

$$\leq \max \text{Prob}\,[\mathbf{H}_2 \subseteq H_1]$$

(since $\mathbf{H}_2$, $\mathbf{H}_1$ are, by definitions, independent)

$$\leq \max\,(1-p)^{\binom{x}{2} - |H_1|} \leq (1-p)^{\frac{1}{2}\binom{x}{2}}$$

$$< \frac{1}{2}\binom{n}{x}^{-1}$$

by elementary calculations.

By (5.8)

$$|\mathbf{H}_1| \leq \sum_{i \notin X} \binom{\mathbf{Y}_i}{2} \tag{5.12}$$

where

$$\mathbf{Y}_i = |\{j \in X : \{i, j\} \in \mathbf{G}\}|. \tag{5.13}$$

The $\mathbf{Y}_i$ are independent with binomial distribution $B(x, p)$. We require a purely probabilistic result.

LEMMA 5.3: $\text{Prob}\left[\sum_{i \notin X} \binom{\mathbf{Y}_i}{2} > \frac{1}{2}\binom{x}{2}\right] < \frac{1}{2}\binom{n}{x}^{-1}.$

PROOF (outline): Intuitively, $\mathsf{E}(\mathbf{Y}_i) = xp \sim 2c_2^{1/2} \log n$ so $\frac{1}{2} \sum \mathbf{Y}_i^2$ should be $\sim \frac{1}{2} n(2c_2^{1/2} \log n)^2 < \frac{1}{4} x^2$ by a factor of c_2. $\frac{1}{2} \sum \mathbf{Y}_i^2$ has a noncentral chi square distribution. Set $z_i = 2^i c_2^{2/3}(\log n)$, $0 \leq i \leq \log_2 n$ and set

$$(5.14) \qquad w_i = \begin{cases} n4^{-i}(i+1)^{-2}, & 0 \leq i \leq \frac{1}{4} \log_2 n \\ n4^{-i} i^{-1}, & \frac{1}{4} \log_2 n < i \leq \log_2 n \,. \end{cases}$$

Then

$$(5.15) \qquad \sum_i \text{Prob}\,[\geq w_i \text{ of the } \mathbf{Y}_j > z_i] < \frac{1}{2} \binom{n}{x}^{-1}.$$

If for all i less than w_i of the $\mathbf{Y}_j > z_i$ then

$$(5.16) \qquad \sum_i \binom{\mathbf{Y}_i}{2} < \sum_i w_i \binom{z_{i+1}}{2} < \frac{1}{2} \binom{x}{2}.$$

We return to the proof of Lemma 5.2.

Combining Lemma 5.3 and (5.12) yields

$$(5.17) \qquad \text{Prob}\left(|\mathbf{H}_1| > \frac{1}{2} \binom{x}{2} \right) < \frac{1}{2} \binom{n}{x}^{-1}.$$

Formulae (5.10), (5.11) and (5.17) yield (5.9) which proves Lemma 5.2. Q.E.D.

The following result is an improvement on the upper bound (5.2).

THEOREM 5.4 (J. E. GRAVER and J. YACKEL [1968]):

$$R(3, y) \leq c_3 y^2 \log \log y / \log y \,.$$

PROOF: It will be more convenient to show

$$R(3, y+1) \leq c_4 y^2 \log \log y / \log y \,.$$

Let $n = R(3, y+1) - 1$. Let G be a graph on $[n]$ containing no triangle and no set of $y+1$ independent points. To avoid a triangle the points adjacent to a given point must be independent. Therefore each point has degree $\leq y$. Fix an independent set $A \in [n]^y$. We know such a set exists because we cannot add an isolated point to G. For $0 \leq i \leq y$ set

$$A_i = \{b \in [n] - A : |\{a \in A : \{a, b\} \in G\}| = i\}$$
$$p_i = |A_i|$$
$$e_i = |G \mid A_i|.$$

So

$$(5.18) \qquad R(3, y+1) = n + 1 = 1 + y + \sum_{i=0}^{y} p_i \,.$$

Since $A \cup A_0$ is an independent set, $A_0 = \varnothing, p_0 = e_0 = 0$. Now suppose $|A_1| > |A|$ (see Fig. 1). By the pigeon-hole principle there exist $b, c \in A_1$, $\{a, b\}, \{a, c\} \in G$. The set $A - \{a\} + \{b, c\}$ is independent with $y + 1$ elements, a contradiction. Thus $p_1 = |A_1| \leq y$. Formula (5.18) then implies

$$R(3, y+1) \leq 1 + 2y + \sum_{i=2}^{y} p_i\,. \tag{5.19}$$

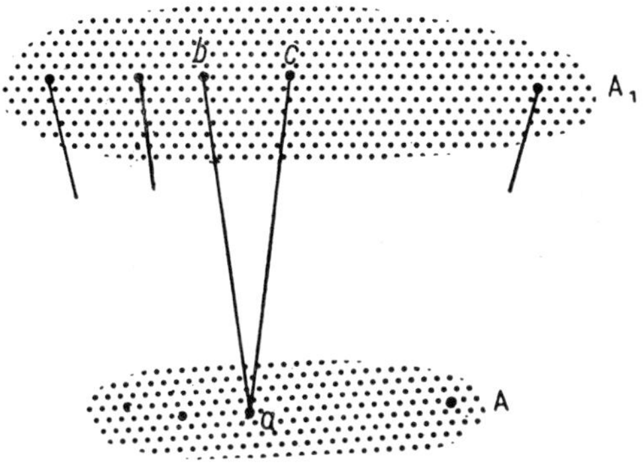

Fig. 1 $[|A_1| > |A|]$

We now count $|G \cap (A \times A^c)|$ in two ways.

$$\begin{aligned} \sum_{i=1}^{y} i p_i &= \sum_{i=1}^{y} \sum_{b \in A_i} i \\ &= \sum_{b \in A^c} |\{a \in A : \{a, b\} \in G\}| \\ &= |G \cap (A \times A^c)| \\ &= \sum_{a \in A} |\{b \in A^c : \{a, b\} \in G\}| \\ &\leq y^2 \end{aligned} \tag{5.20}$$

since each $a \in A$ has degree $\leq y$. In particular, this implies

$$\sum_{i=h}^{y} p_i \leq y^2/h\,. \tag{5.21}$$

Setting $h = 2$,

$$R(3, y+1) \leq 1 + 2y + y^2/2\,. \tag{5.22}$$

To improve (5.22) we bound p_i for small $i \geq 2$.

Lemma 5.4: *For* $2 \le i \le y$, $p_i \le 2y^{2-1/i}$.

Proof: We want to "transfer" points of A so that we get a larger independent set, as with $i = 1$. For each $b \in A_i$ set $S_b = \{a \in A : \{a, b\} \in G\}$, so $|S_b| = i$. Let $\mathbf{A}$ be a random subset of A with each $a \in A$ having independent Prob $[a \in \mathbf{A}] = y^{-1/i}$. Set $X(\mathbf{A}) = \{b \in A_i : S_b \subseteq \mathbf{A}\}$. We have

$$\text{(5.23)} \qquad \begin{aligned} E(|\mathbf{A}|) &= y^{1-1/i} \\ E(|X(\mathbf{A})|) &= p_i[y^{-1/i}]^i = p_i y^{-1}. \end{aligned}$$

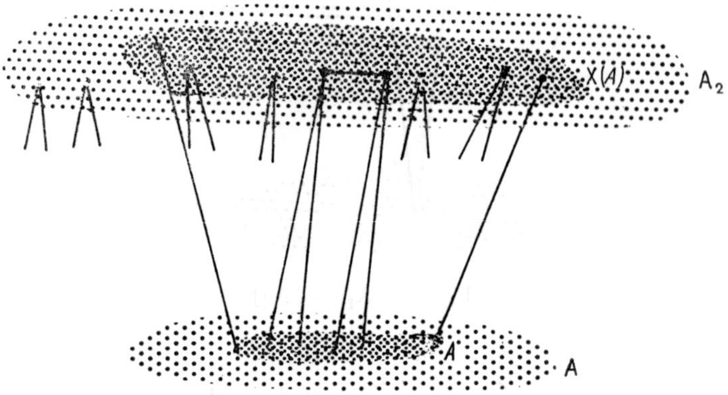

Fig. 2 [i = 2]

Because there may be edges in $X(\mathbf{A})$, the set $A - \mathbf{A} + X(\mathbf{A})$ isn't necessarily independent (see Fig. 2). If $\{b, c\} \in G \mid X(\mathbf{A})$ then $S_b \cup S_c \subseteq A$. Since G contains no triangles $S_b \cap S_c = \varnothing$. Thus

$$\text{(5.24)} \qquad \text{Prob}\,[S_b \cup S_c \subseteq \mathbf{A}] = (y^{-1/i})^{2i} = y^{-2}.$$

Therefore

$$\text{(5.25)} \qquad E(|G \mid X(\mathbf{A})|) = e_i y^{-2}.$$

Since each point of A_i has degree $\le y$, $e_i \le \frac{1}{2} p_i y$. Therefore, (5.25) implies

$$\text{(5.26)} \qquad E(|G \mid X(\mathbf{A})|) \le \frac{1}{2} p_i y^{-1}.$$

It is trivial to prove that if a graph has n points and e edges it contains an independent set of $n - e$ points. Thus there exists an independent $Y(\mathbf{A}) \subseteq X(\mathbf{A})$, $|Y(\mathbf{A})| \ge |X(\mathbf{A})| - |G \mid X(\mathbf{A})|$. Then $A - \mathbf{A} + Y(\mathbf{A})$

is independent and

$$(5.27)\qquad \mathsf{E}[|A - \mathbf{A} + Y(\mathbf{A})|] \geq |A| - \mathsf{E}(\mathbf{A}) + \mathsf{E}(|X(\mathbf{A})|) - \mathsf{E}(|G|X(\mathbf{A})|) \geq y - y^{1-1/i} + \frac{1}{2}p_iy^{-1}.$$

If $\frac{1}{2}p_iy^{-1} > y^{1-1/i}$ there exists a value A^* of $\mathbf{A}$ so that $|A - A^* + Y(A^*)| > y$, a contradiction. Thus $p_i \leq 2y^{2-1/i}$, which completes Lemma 5.4.

From Lemma 5.4, (5.19) and (5.21),

$$(5.28)\qquad R(3, y+1) \leq 1 + 2y + \sum_{i=2}^{h-1} 2y^{2-1/i} + y^2/h.$$

Setting $h = \log y/\log\log y$ yields Theorem 5.4. Q.E.D.

Corollary 5.1 (J. Yackel [1972]): *For* $3 \leq x \leq y$

$$R(x, y) \leq c_5 \frac{\log\log y}{\log y}\binom{x+y-2}{y-1}.$$

We suppress the difficult, albeit elementary, proof which depends only on Theorem 5.4 and (5.1).

Exercises

1. Combining (5.2) and Theorem 5.1 yields

$$\sqrt{2} \leq \underline{\lim}\, R(k, k)^{1/k} \leq \overline{\lim}\, R(k, k)^{1/k} \leq 4.$$

a) Prove, or disprove, that $\lim_k R(k, k)^{1/k}$ exists. ($25).

b) Find $\lim_k R(k, k)^{1/k}$ if it exists. ($25).

2. Let $k \geq 3$ be fixed and $n \to \infty$. Prove there exists a graph G on n points containing no k-clique and no set of $[C_kn^{2/(k-1)}\log n]$ independent points. Show that this implies

$$R(k, n) \geq C'_k\,[n/\log n]^{(k-1)/2}.$$

3. (?) Let $\varepsilon > 0$, $k > 3$. Prove or disprove that for n sufficiently large

$$R(k, n) \geq n^{k-1-\varepsilon}.$$

4. (Hard) Set $\mathbf{G}' = \mathbf{G}_{n,p} - \{\{i,j\} : \exists\, k\{i,j,k\}^2 \subseteq G_{n,p}\}$ where $p = 2c_2^{-1/2}n^{-1/2}$. That is, $\mathbf{G}'$ is $\mathbf{G}_{n,p}$ with all triangles deleted. Show that for c_2 sufficiently large Prob [$\mathbf{G}'$ contains x independent points] < 1 where $x = [c_2\sqrt{n}\log n]$.

5. Let c be a coloring of $[n]^2$. We consider c a map from the ordered pairs (x, y), $x < y$, into the set of colors. We call c canonical on $A \subseteq [n]$ if one of the following is true for $x_i, y_i \in A$.

(a) $c(x_1, x_2) = c(y_1, y_2)$ iff $x_1 = y_1$ and $x_2 = y_2$
(b) $c(x_1, x_2) = c(y_1, y_2)$ iff $x_1 = y_1$
(c) $c(x_1, x_2) = c(y_1, y_2)$ iff $x_2 = y_2$
(d) $c(x_1, x_2) = c(y_1, y_2)$ always

Denote by

$$(*) \quad n \to (k)^2_{\text{can}}$$

the statement that any coloring (with any number of colors) of $[n]$ is canonical on some $A \in [n]^k$. P. Erdős and R. Rado [1956] showed that for all k there exists n such that (*) holds. Prove (F. Galvin, private correspondence) that if

$$n < \left[\frac{k}{\sqrt{2} + 0(1)}\right]^k$$

then (*) does not hold. $\left(\text{Hint: color } [n]^2 \text{ randomly with } \binom{k}{2} - 1 \text{ colors.}\right)$

6. VAN DER WAERDEN'S THEOREM

VAN DER WAERDEN'S THEOREM [1925]: *Given an integer t there exists a minimal integer $m = m(t)$ such that if $\{1, 2, \ldots, m\}$ is 2-colored there exists a monochromatic arithmetic progression of length t.*

The proof of VAN DER WAERDEN's Theorem yields as an upper bound to $m(t)$ a function $m_1(t)$ that increases extremely rapidly. No "reasonable" upper bound on $m(t)$ has been found. We shall be solely concerned with lower bounds for $m(t)$.

Set $A(a, \delta, t) = \{a, a + \delta, \ldots, a + (t-1)\delta\}$. For m, t fixed set $\mathfrak{J} = \{A(a, \delta, t) : \delta > 0, 1 \leq a < a + (t-1)\delta \leq m\}$ = the set of arithmetic progressions of length t contained in $[m]$. Then $|\mathfrak{J}| \leq m^2/2t$ and $\mathfrak{J}$ is a t-graph. If $m \geq m(t)$ then $\mathfrak{J}$ does not have property B. Thus, by Theorem 4.2,

THEOREM 6.1 (P. ERDŐS and R. RADO [1952]):

$$m(t) \geq \sqrt{2t2^t}\,(1 - o(1)).$$

This result was improved by W. M. SCHMIDT [1962]. Let m, t be fixed. Set $u = [(t/\log t)^{1/2}]$. Let $C = (K_1, K_2)$ be a coloring of $\{1, \ldots, m\}$. We attempt to adjust C slightly to destroy the monochromatic arithmetic progressions. If P is a progression of length t and $i = 1$ or 2 we call P "near i" if $|P \cap K_i| > t - u$. For each "near i" P pick $x(P) \in P \cap K_i$ in any arbitrary manner. Set $C^* = (K_1 \Delta N, K_2 \Delta N)$ where

$$N = \{x(P) : P \text{ is near } i \text{ for } i = 1 \text{ or } 2\}.$$

If C^* contains a monochromatic arithmetic progression Q (say colored i) then Q could not have been "near i" in C. Thus there would exist $i = 1$ or 2, r, $l \geq r \geq u + 1$, progressions $P_1, \ldots, P_r$, $Q = \{q_1, \ldots, q_t\}$ such that (in C)

1. Each P_j is "near i".
2. $q_j \in P_j \cap K_i$ for $1 \leq j \leq r$.
3. $q_j \notin K_i$ for $r < j \leq l$.

By a complex series of lemmas, W. M. SCHMIDT shows that for

$$m < 2^{t - c(t \log t)^{1/2}}$$

a random **C** would not have this property and therefore $\mathbf{C}^*$ would not contain monochromatic arithmetic progressions of length t. Thus

THEOREM 6.2 (W. M. SCHMIDT [1962]): $m(t) > 2^{t-c(t\log t)^{1/2}}$.

This section, however, must be considered a victory for the constructivists.

THEOREM 6.3 (E. R. BERLEKAMP [1968]): *If t is a prime*

$$m(t) \geq t2^t.$$

BERLEKAMP's proof is completely constructive. It is based on the Galois field $GF(2^t)$. Since it is not known if there always exists a prime p, $t - \sqrt{t} < p < t$, we cannot say that BERLEKAMP's result is an improvement of SCHMIDT's result for all sufficiently large t.

Exercises

1. Let $W(k, t)$ denote the least integer m such that every k-coloration of $\{1, \ldots, t\}$ contains a monochromatic arithmetic progression of length t. Prove that

$$W(k, t) \geq (2tk^{t-1})^{1/2}(1 - o(1)).$$

(Note: L. MOSER [1960] has shown $W(k, t) > tk^{c \log k}$ by constructive methods. For t fixed, $k \to \infty$; this is an improvement of Exercise 1.)

7. QUASI-RAMSEY THEOREMS

RAMSEY's Theorem is concerned with finding large monochromatic subgraphs in an edge-coloring of the complete graph K_n. In this section we look for subgraphs that are predominantly one-color.

We denote a 2-coloring of K_n by h. It will be convenient to think of h as a function with domain $[n]^2$ and range $\{+1, -1\}$. We induce a function, also denoted by h, defined on the subsets of $[n]$ by

$$h(S) = \sum_{A \in [S]^2} h(A) . \tag{7.1}$$

Then $|h(S)|$ is a measure of the discrepancy in the coloring of h. Set

$$g(n, t) = \min_{n} \max_{\substack{S \subseteq [n] \\ |S| \leq t}} |h(S)| . \tag{7.2}$$

THEOREM 7.1: *Let $n \geq t$ be sufficiently large.*

$$\text{If } t \leq \frac{\log n}{2 \log 2} \quad g(n, t) = \binom{t}{2} .$$

(7.3)

$$\text{If } t > \frac{\log n}{2 \log 2}$$

$$10^{-3} t^{3/2} \sqrt{\log (5n/t)} \leq g(n, t) \leq t^{3/2} \sqrt{\log (5n/t)} . \tag{7.4}$$

PROOF: If $t \leq \dfrac{\log n}{2 \log 2}$ then, by (5.2), $R(t, t) \leq \dbinom{2t-2}{t-1} \leq n$ so $g(n, t) = \dbinom{t}{2}$. Fix $t > \dfrac{\log n}{2 \log 2}$ and let $\mathbf{h}$ be a random coloring. If $|S| = s < t$, $\mathbf{h}(S) \sim \mathbf{S}_{\binom{s}{2}}$, as defined by (3.1). Thus

$$\text{Prob}\,[\mathbf{h}(S) > \alpha] < e^{-\alpha^2/2\binom{s}{2}} < e^{-\alpha^2/t^2} . \tag{7.5}$$

So

$$\text{Prob}\,[\exists S, |S| \leq t, \mathbf{h}(S) > \alpha] < n^t e^{-\alpha^2/t^2} < 1 \tag{7.6}$$

for

$$\alpha = t^{3/2} \sqrt{\log (5n/t)} . \tag{7.7}$$

Thus

$$g(n, t) \leq t^{3/2} \sqrt{\log (5n/t)} . \tag{7.8}$$

We induce a function, also denoted by h, defined on pairs of disjoint subsets of $[n]$ by

$$h(ST) = \sum_{\substack{x \in S \\ y \in T}} h(\{x, y\}) . \tag{7.9}$$

We now show the lower bound (7.4). Assume $t \leq n/2$. Assume, for convenience, $n \equiv t \equiv 0(2)$. We first find disjoint X, B, $| X \cup B | \leq t$, $h(XB)$ large.

Decompose $[n] = A_1 + A_2$, $| A_1 | = | A_2 | = n/2$. Let $\mathbf{B} \subseteq A_2$ be defined by

$$\text{Prob}\, [y \in \mathbf{B}] = p = t/4n \quad \text{for} \quad y \in A_2 . \tag{7.10}$$

Fix $x \in A_2$. Then

$$h(x\mathbf{B}) \approx B(n_x, p) - B\left(\frac{n}{2} - n_x, p\right) \tag{7.11}$$

where $n_x = | \{y \in A_2 : h(\{x, y\}) = 1\} |$. Set $\alpha = 10^{-1} t^{1/2} \sqrt{\log (5n/t)}$. By elementary probabilistic calculations

$$\text{Prob}\, [| h(x\mathbf{B}) | > \alpha] > t/n . \tag{7.12}$$

Set

$$X(\mathbf{B}) = | \{x \in A_1 : | h(x\mathbf{B}) | > \alpha\} | . \tag{7.13}$$

Then

$$\mathsf{E}[| X(\mathbf{B}) |] = \sum_{x \in A_1} \text{Prob}\, [| h(x\mathbf{B}) | > \alpha] > t/2 . \tag{7.14}$$

Since $| X(\mathbf{B}) | \leq n/2$,

$$\text{Prob}\, [| X(\mathbf{B}) | \geq t/4] \geq t/2n . \tag{7.15}$$

By elementary calculations,

$$\text{Prob}\, [| \mathbf{B} | \leq t/2] > 1 - 1/n . \tag{7.16}$$

Combining (7.15), (7.16) there exists $B \subseteq A_2$, $| B | \leq t/2$, $| X(B) | \geq t/4$. Then either

$$| \{x \in A_1 : h(xB) > \alpha\} | \geq t/8$$

or

$$| \{x \in A_1 : h(xB) < -\alpha\} | \geq t/8 .$$

By symmetry (between h and $-h$) we assume the former. Fix

$$X \subseteq \{x \in A_1 : h(xB) > \alpha\}, \quad |X| = [t/8].$$

Then

$$h(XB) = \sum_{x \in X} h(xB) > |X| \alpha > 10^{-2} t^{3/2} \log \sqrt{(5n/t)}. \tag{7.17}$$

We translate the discrepancy (7.17) into a discrepancy of the form $h(S)$. Note that

$$h(X) + h(B) + h(XB) = h(X \cup B). \tag{7.18}$$

Therefore

$$g(n, t) \geq h(S) \geq \frac{h(XB)}{3} \geq (300)^{-1} t^{3/2} \sqrt{\log (5n/t)} \tag{7.19}$$

where S is one of the sets X, B, $X \cup B$.

We have assumed $t \leq n/2$. For $t > n/2$

$$g(n, t) \geq g(n, n/2) \geq 10^{-3} t^{3/2} \sqrt{\log (5n/t)} \tag{7.20}$$

Q.E.D.

COROLLARY 7.1: *Let* $g(n) = \min_h \max_{S \subseteq [n]} |h(S)|$. *Then*

$$a_1 n^{3/2} \leq g(n) \leq a_2 n^{3/2}$$

where a_1, a_2 *are positive absolute constants.*

COROLLARY 7.2: *For* $1 \geq \varepsilon > 0$ *let*

$$g(n, \varepsilon) = \min_h \max \left\{ |S| : S \subseteq [n], |h(S)| \geq \varepsilon \binom{|S|}{2} \right\}.$$

Then for $n > n(\varepsilon)$

$$b_1 \varepsilon^{-2} \log n \leq g(n, \varepsilon) \leq b_2 \varepsilon^{-2} \log n$$

where b_1, b_2 *are positive absolute constants.*

Theorem 7.1 may be generalized to k-graphs. Let k be fixed, $n > n(k)$. Let $h = h_k$ be a 2-coloring of the complete k-graph on $[n]$. That is, $h : [n]^k \to \{+1, -1\}$. We induce a function, also denoted by h, defined on the subsets of $[n]$ by

$$h(S) = \sum_{A \in [S]^k} h(A). \tag{7.21}$$

$$g_k(n, t) = \min_h \max_{\substack{S \subseteq [n] \\ |S| \leq t}} |h(S)|. \tag{7.22}$$

THEOREM 7.2: *For* $t \leq (\log n)^{1/(k-1)}$

$$c_{k1} \binom{t}{k} \leq g_k(n, t) \leq \binom{t}{k}. \tag{7.23}$$

For $n \geq t \geq (\log n)^{1/(k-1)}$

(7.24) $$c_{k2} t^{(k+1)/2} \sqrt{\log (5n/t)} \leq g_k(n, t) \leq c_{k3} t^{(k+1)/2} \sqrt{\log (5n/t)}\,.$$

The upper bound (7.23) is trivial. The lower bound (7.23) is Theorem 12.3. The proof of (7.24) is left as Exercises 1 and 2.

Exercises

1. Prove the upper bound (7.24).
2. (Difficult) Read the proof (P. ERDŐS and J. SPENCER [1972]) of the lower bound (7.24) for $t = n$. Prove the lower bound (7.24) for all t.
3. Let $|A_1| = |A_2| = n$, $A_1 \cap A_2 = \emptyset$. Set

$$f(n) = \min_{h : A_1 A_2 \to \{+1, -1\}} \max_{B_i \subseteq A_i} |h(B_1 B_2)|$$

Prove

$$c_1 n^{3/2} > f(n) > \frac{n^{3/2}}{2\sqrt{\pi}}(1 - o(1))\,.$$

8. A QUASI VAN DER WAERDEN THEOREM

We represent a 2-coloring of $[n]$ by a function $h : [n] \to \{+1, -1\}$. Analogously to the previous section, we look for arithmetic sequences that are predominantly one-color. More precisely, set

$$G(n) = \min_{h} \max_{a,d,m} \left| \sum_{u=0}^{m} h(a+ud) \right| \quad (0 < a < a+dm \leq n). \tag{8.1}$$

K. F. Roth [1964] showed $G(n) > n^{1/4}$ for n sufficiently large. Given h, he defined

$$S(\alpha) = \sum_{j=1}^{n} h(j)\, e^{2\pi i j\alpha}, \quad i = \sqrt{-1} \tag{8.2.}$$

and employed the theory of complex variables. We shall find an upper bound to $G(n)$. We use probabilistic methods to first obtain a rough bound and then improve it.

Let $\mathbf{h}$ be a random coloring.

$$\sum_{u=0}^{m} \mathbf{h}(a+ud) \sim \mathbf{S}_{m+1}\,. \tag{8.3}$$

Thus

$$\mathrm{Prob}\left[\left|\sum_{u=0}^{m} \mathbf{h}(a+ud)\right| > \alpha\right] < e^{-\alpha^2/2(m+1)}. \tag{8.4}$$

As $m+1 \leq n$ this probability is $< n^{-3}$ for $\alpha = c\sqrt{n}\log n$. As there are $< n^3$ choices for a, d, m,

$$\mathrm{Prob}\left[\max_{a,d,m}\left|\sum_{u=0}^{m} \mathbf{h}(a+ud)\right| > \alpha\right] < 1 \tag{8.5}$$

and therefore

$$G(n) < c\sqrt{n \log n}\,. \tag{8.6}$$

How may we improve this result? Let $1 \leq a \leq d$. The sums $\sum_{u=i}^{j} \mathbf{h}(a+ud)$ are strongly dependent. In fact

$$\mathrm{Prob}\left[\exists\, i, j \left|\sum_{u=i}^{j} \mathbf{h}(a+ud)\right| > \alpha\right]$$

$$
\begin{aligned}
&< \text{Prob}\left[\exists j \;\left|\sum_{u=0}^{j} \mathbf{h}(a+ud)\right| > \alpha/2\right] \\
(8.7)\quad &< 2\,\text{Prob}\left[\left|\sum_{u=0}^{[(n-a)/d]} \mathbf{h}(a+u\,d)\right| > \alpha/2\right] \\
&< 2e^{-d\alpha^2/8(n+d)}.
\end{aligned}
$$

The first inequality is elementary. The second is due to A. N. Kolmogoroff.* The third inequality is based on (2.2) and $1 + [(n-a)/d] < (n+d)/d$. Since for each d there are d choices of a, $1 \leq a \leq d$,

$$
\begin{aligned}
&\text{Prob}\left[\exists\, a, d, m \;\left|\sum_{u=0}^{m} \mathbf{h}(a+ud)\right| > \alpha\right] \\
(8.8)\quad &< \sum_{d=1}^{n} 2de^{-d\alpha^2/8(n+d)} \\
&< 1
\end{aligned}
$$

for $\alpha = 10\sqrt{n}$ by a simple calculation. Thus $G(n) \leq 10\sqrt{n}$.

There is a further improvement (J. Spencer [1971*b*]). The sequences most likely to have high discrepancy have d small. Suppose $h(2i-1) = -h(2i)$ for $1 \leq i \leq [n/2]$. Then all sequences with $d = 1$ would have discrepancy 0, +1, −1. Some calculation reveals an appropriate generalization of this. Set $k = [(\log n)/4]$, $N = gcd\ (1, \ldots, k)$. By the Prime-Number Theorem $N \sim e^k < n^{1/3}$. (Actually, we could take $N = k!$ with similar results.) Let

$$(8.9)\quad \mathcal{H} = \{h : 0 \leq j < N \text{ implies } h\big((2i+1)N+j\big) = -h(2i\,N+j)\}.$$

That is, $[n]$ is split into blocks of $2N$ with the second half of the block negating the first half. Let $\mathbf{h}$ be a random element of $\mathcal{H}$. If $d \mid N$ then for any $h \in \mathcal{H}$

$$(8.10)\quad \left|\sum_{u=0}^{m} h(a+ud)\right| < 2N/d < 2n^{1/3}.$$

If $d \nmid N$ and $1 \leq a \leq d$ the sequence $a, a+d, \ldots$ contains no two elements differing by N. Therefore $\mathbf{h}$ is a random coloring of this

* The general result is

$$\text{Prob}\,[\mathbf{S}_m \geq \alpha] \geq \frac{1}{2}\,\text{Prob}\,[\mathbf{S}_i \geq \alpha \text{ for some } 1 \leq i \leq m].$$

This may be shown by letting w_i = "$\mathbf{S}_j < \alpha$, $1 \leq j < i$, $\mathbf{S}_i \geq \alpha$" and noting the w_i decomposes the RHS and

$$\text{Prob}\,[\mathbf{S}_m \geq \alpha \mid w_i] = \text{Prob}\,[\mathbf{S}_m \geq \alpha \mid \mathbf{S}_i \geq \alpha] \geq \frac{1}{2}.$$

sequence and

$$\text{Prob}\left[\exists\, ij \;\middle|\; \left|\sum_{u=i}^{j} \mathbf{h}(a+ud)\right| > \alpha\right] < 2e^{-d\alpha^2/8(n+d)} \tag{8.11}$$

as before. Setting $\alpha = 100\sqrt{n}\sqrt{\log\log n}/\sqrt{\log n}$,

$$\text{Prob}\left[\exists\, a, d, m \;\middle|\; \left|\sum_{u=0}^{m} \mathbf{h}(a+ud)\right| > \alpha\right] < \sum_{d>k} 2e^{-d\alpha^2/8(n+d)} < 1\,. \tag{8.12}$$

This implies

THEOREM 8.1: $n^{1/4} \leq G(n) \leq 100\sqrt{n}\sqrt{\log\log n}/\sqrt{\log n}$.

It is conjectured that for all $\varepsilon > 0$, $G(n) > n^{1/2-\varepsilon}$ for n sufficiently large.

The following results were discovered as the book was going to press.

This conjecture was recently destroyed by the following observation of A. SÁRKÖZI. Let p be a large prime. Let Z_p denote the ring of integers modulo p. We may find an $h\colon Z_p \to \{-1, +1\}$ such that

(a) $\sum_{\substack{i\in Z_p \\ i\neq o}} h(i) = o$

(b) $|\sum_{i\in A} h(i)| < c\sqrt{p}\sqrt{\log p}$ where A is any arithmetic progression in Z_p.

Such an h may be found by nonconstructive means. One may also set $h(0)$ arbitrarily and, for $i \neq 0$, $h(i) = \left(\frac{i}{p}\right)$, the Legendre symbol. Then, by an analytic number theory result, (b) holds with $\sqrt{p}\sqrt{\log p}$ replaced by $\sqrt{p}\log p$.

Now suppose $n = [p^{3/2}\sqrt{\log p}]$. Define $h^*\colon [n] \to \{-1, +1\}$ by $h^*(j + Kp) = h(j)$, $o \leq j < p$, K integral. The maximal discrepancy can be shown to be $\sqrt{p}\sqrt{\log p}$. So $H(n) \leq \sqrt{p}\sqrt{\log p} = 0\big((n\log n)^{1/3}\big)$.

9. TOURNAMENTS

Tournaments are defined in Section 2. We prove two theorems on tournaments in Section 1. J. W. MOON's [1968] monograph, *Topics on Tournaments*, contains these and many other results on tournaments.

We say that a tournament has property $S(k)$ if for every set A of k players some other player beats all $x \in A$. Note the difficulty of determining the top k players in such a tournament. Set $s(k) =$ the minimal number of players in a tournament with property $S(k)$. K. SCHÜTTE asked to determine $s(k)$ if it existed. By constructive methods E. and G. SZEKERES [1965] showed $s(k) \geq 2^k(k+1) - 1$.

THEOREM 9.1 (P. ERDŐS [1963]):

$$s(k) \leq 2^k k^2(\log 2 + o(1)).$$

PROOF: Let $\mathbf{T} = \mathbf{T}_n$ be a random tournament. Fix $A \in [n]^k$. For each $y \notin A$

$$\text{Prob}\,[\exists\, a \in A,\ (y, a) \notin \mathbf{T}] = 1 - 2^{-k}. \tag{9.1}$$

These probabilities are independent for $y \notin A$. Hence

$$\text{Prob}\,[\forall y \notin A\ \exists\, a \in A,\ (y, a) \notin \mathbf{T}] = (1 - 2^{-k})^{n-k}. \tag{9.2}$$

Therefore

$$\begin{aligned} \text{Prob}\,[\text{not } S(k)] &= \text{Prob}\,[\exists A \in [n]^k\ \forall y \notin A\ \exists\, a \in A,\ (y, a) \notin \mathbf{T}] \\ &\leq \binom{n}{k} (1 - 2^{-k})^{n-k} < 1 \end{aligned} \tag{9.3}$$

for $n = 2^k k^2(\log 2 + o(1))$. For this n

$$\text{Prob}\,[\mathbf{T} \text{ has } S(k)] > 0 \tag{9.4}$$

and therefore $s(k) \leq n$. Q.E.D.

Given any tournament $T = T_n$, it is natural to ask for a ranking of the players. A ranking is given by a permutation σ on $\{1, 2, ..., n\}$.

(Here $\sigma(x)$ = rank of x.) This induces a transitive tournament $P = P_\sigma$ on $\{1, 2, \ldots, n\}$ by $(x, y) \in P$ iff $\sigma(x) < \sigma(y)$. An index of the quality of the ranking is $|P \cap T|$. Given T we may wish to find P that maximizies $|P \cap T|$. Set

$$f(n) = \min_{T_n} \max_{P_i} |P \cap T| . \tag{9.5}$$

The definition is due to P. ERDŐS and J. W. MOON [1965]. They called $P \cap T$ a consistent set of arcs. For any T and any P

$$|P \cap T| + |P^c \cap T| = \binom{n}{2}$$

(P^c is P with arcs reversed) so $f(n) \geq \frac{1}{2}\binom{n}{2}$.

THEOREM 9.2 (J. SPENCER [1971*a*]):

$$\frac{1}{2}\binom{n}{2} + c_1 n^{3/2} \leq f(n) \leq \frac{1}{2}\binom{n}{2} + \left(\frac{\sqrt{2}}{4} + o(1)\right) n^{3/2}\sqrt{\log n}$$

where c_1 is a positive absolute constant.

PROOF: Both bounds are by probabilistic methods. For the upper bound let $\mathbf{T} = \mathbf{T}_n$ be random. For fixed P

$$|\mathbf{T} \cap P| \sim \mathbf{U}_{\binom{n}{2}} \sim \frac{1}{2}\mathbf{S}_{\binom{n}{2}} + \frac{1}{2}\binom{n}{2} \tag{9.6}$$

where $\mathbf{U}$, $\mathbf{S}$ are defined by (3.1), (3.2)

$$\begin{aligned} &\operatorname{Prob}\left[|\mathbf{T} \cap P| \geq \frac{1}{2}\binom{n}{2} + \alpha\right] \\ &= \operatorname{Prob}\left[\mathbf{S}_{\binom{n}{2}} \geq 2\alpha\right] < e^{-4\alpha^2/n^2} \end{aligned} \tag{9.7}$$

by (3.9). Thus

$$\operatorname{Prob}\left[\exists P, |\mathbf{T} \cap P| \geq \frac{1}{2}\binom{n}{2} + \alpha\right] < n!\, e^{-4\alpha^2/n^2} < 1 \tag{9.8}$$

for $\alpha = \frac{1}{2} n^{3/2}\sqrt{\log n}$.

We improve this by taking advantage of the strong dependence of $\mathbf{T} \cap P_1$ and $\mathbf{T} \cap P_2$ when P_1 and P_2 are "close". Let us, for convenience, assume $n = m^2$. Set

$$\mathscr{P} = \{P = P_\sigma : \sigma(im + 1) < \sigma(im + 2) < \ldots < \sigma(im + m) \text{ for } 0 \leq i \leq m - 1\} . \tag{9.9}$$

Then

$$(9.10) \qquad |\mathcal{S}| = \frac{n!}{m!^m} < n^{\frac{n}{2}(1+o(1))}$$

so

$$(9.11) \quad \text{Prob}\left[\exists P \in \mathcal{S}, |\mathbf{T} \cap P| \geq \frac{1}{2}\binom{n}{2} + \alpha\right] < n^{\frac{n}{2}(1+o(1))} e^{-4\alpha^2/n^2!} < 1$$

for $\alpha = \left(\frac{\sqrt{2}}{4} + o(1)\right) n^{3/2}\sqrt{\log n}$. Fix T, $|T \cap P| \leq \frac{1}{2}\binom{n}{2} + \alpha$ for all $P \in \mathcal{S}$. Let σ be any permutation. Define σ' such that for $1 \leq i \leq m$

$$\{\sigma(im+1), \ldots, \sigma(im+m)\} = \{\sigma'(im+1), \ldots, \sigma'(im+m)\}$$

and

$$\sigma'(im+1) < \ldots < \sigma'(im+m).$$

Then $P_{\sigma'} \in \mathrm{P}$ and

$$|P_\sigma \mathbin{\Delta} \mathrm{P}_{\sigma'}| \leq m\binom{m}{2} \leq n^{3/2}/2.$$

Thus, for all σ

$$|T \cap P_\sigma| \leq |T \cap P_{\sigma'}| + |P_\sigma \mathbin{\Delta} P_{\sigma'}| \leq \alpha + n^{3/2}/2$$

$$(9.12) \qquad \leq \left(\frac{\sqrt{2}}{4} + o(1)\right) n^{3/2}\sqrt{\log n}.$$

The lower bound follows from our Quasi-Ramsey Theorem. Fix $T = T_n$ and define $h : [n]^2 \to \{+1, -1\}$ by $h(\{x, y\}) = 1$ if $(x, y) \in T$ and $x < y$, -1 otherwise. By Corollary 7.1 there exists $B \subseteq [n]$, $|h(B)| \geq c_1 n^{3/2}$. Set $B^c = [n] - B$. If $h(B) \geq c_1 n^{3/2}$ rank B by $<$. If $h(B) \leq -c_1 n^{3/2}$ rank B by $>$. If $h(B^c) \geq 0$ rank B^c by $<$, otherwise by $>$. If $h(B \times B^c) > 0$ rank $B < B^c$, otherwise rank $B^c < B$. Letting σ denote the final ranking $|T \cap P_\sigma| \geq \frac{1}{2}\binom{n}{2} + c_1 n^{3/2}$. Q.E.D.

The gap between the bounds of Theorem 9.2 seems difficult to eliminate. Even Exercise 3 is not known.

A k-cycle in a tournament T is a set of arcs of the form $\{(x_1, x_2), (x_2, x_3), \ldots, (x_{k-1}, x_k), (x_k, x_1)\} \subseteq T$ where $x_1, \ldots, x_k$ are distinct nodes. Let $C_k(T)$ denote the number of k-cycles in T, $C_k(n) = \max C(T_n)$ over all tournaments T_n on n players. Theorem 1.2 gives a bound on $C_n(n)$. We now consider k fixed and $n \to \infty$.

THEOREM 9.3: *For k fixed, $n \geq k$, $C_k(n) \Big/ \binom{n}{k}$ is monotone decreasing in n.*

PROOF: Fix $n < m$, T_m such that $C_k(T_m) = C_k(m)$. Let $\mathbf{A}$ be a random member of $[m]^n$. Then

$$\mathsf{E}[C_k(T_m \mid \mathbf{A})] = \sum \text{Prob}\,[\{x_1, \ldots, x_k\} \subseteq \mathbf{A}] \tag{9.13}$$

(the summation over all k-cycles $\{(x_1, x_2), \ldots, (x_k, x_1)\} \subseteq T_m$)
$= C_k(T_m) \binom{n}{k} \Big/ \binom{m}{k}$.

Therefore there exists $A \in [m]^n$

$$C_k(n) \geqq C_k(T \mid A) \geqq \mathsf{E}[C_k(T \mid \mathbf{A})] = C_k(m) \binom{n}{k} \Big/ \binom{m}{k}. \tag{9.14}$$

Q.E.D.

THEOREM 9.4: $C_k(n) \geqq \binom{n}{k} (k-1)!\, 2^{-k}$.

We leave the proof of Theorem 9.4 as Exercise 5. Theorems 9.3, 9.4 imply that

$$\alpha_k = \lim_{n\to\infty} C_k(n)\, n^{-k} \tag{9.15}$$

exists.

For $k = 3$ many authors (see J. W. MOON [1968] p. 9) have shown

$$C_3(n) = \frac{1}{24}(n^3 - n) \quad \text{if } n \text{ is odd} \tag{9.16}$$

$$\frac{1}{24}(n^3 - 4n) \quad \text{if } n \text{ is even.}$$

For $k = 4$, U. COLOMBO [1964], L. W. BEINEKE and F. HARARY [1965] (see J. W. MOON [1968] p. 9) have shown

$$C_4(n) = \frac{1}{48} n(n+1)(n-1)(n-3) \quad \text{if } n \text{ is odd} \tag{9.17}$$

$$\frac{1}{48} n(n+2)(n-2)(n-3) \quad \text{if } n \text{ is even.}$$

Thus $\alpha_4 = \frac{1}{48}$. The probabilistic method, Theorem 9.4, gives only $\alpha_4 \geqq \frac{1}{96}$. The evaluation of α_k, $k \geqq 5$, is an open problem.

Exercises

1. $T = T_v$ is transitive iff there exists a permutation σ on the players so that $|T \cap P_\sigma| = \binom{v}{2}$. For $1/2 < p \leq 1$ say that $T = T_v$ is p-transitive if σ exists so that

$$|T \cap P_\sigma| \geq p\binom{v}{2}.$$

Set $v = v_p(n) =$ the largest integer such that every T_n contains a p-transitive subtournament T_v. For p fixed, find bounds on $v_p(n)$ as n approaches infinity.

2. Say that T has property $S^*(k)$ if for every set A of k players, and every $B \subseteq A$, there exists $x \notin A$ so that x beats all $b \in B$ and is beaten by all $a \in A - B$. Set $s^*(k) =$ the minimal number of players in a tournament with property $S^*(k)$. Show

$$s^*(k) \leq 2^k k^2 (\log 2 + o(1)).$$

3. (?) Set $h(T) = \max_\sigma |T \cap P_\sigma|$. Does

$$\lim_{n\to\infty} \left[\mathrm{E}(h(\mathbf{T}_n)) - \frac{1}{2}\binom{n}{2} \right] n^{-3/2} = \infty\,?$$

4. Let F be a family of tournaments on k players. Let $C(T, F)$ be the number of subtournaments of T in F, $C(n, F) = \max C(T_n, F)$. Prove $C(n, F) \Big/ \binom{n}{k}$ is monotone decreasing in n.

5. Prove Theorem 9.4. (Hint: Find $\mathrm{E}[C_k(\mathbf{T}_n)]$).

10. REGULAR TOURNAMENTS

We define a partial order P as a set of ordered pairs (x, y), $x \neq y$ such that $(x, y), (y, z) \in P$ implies $(x, z) \in P$. Note that a subset of a partial order is not necessarily a partial order. If A, B are disjoint then all subsets of $A \times B$ are partial orders. R. C. ENTRINGER, P. ERDŐS and C. HARNER [to appear] asked for the maximal integer $q = q(n)$ such that every tournament $T = T_n$ has some partial order $P \subseteq T$ with $|P| \geq q$. In this section we show

THEOREM 10.1: $\frac{n^2}{8} + \left(\frac{1}{2} - o(1)\right) \log_2 n \leq q(n) \leq \frac{n^2}{8} + c_1 \log_2 n.$

The proof shall be sketchy. We refer to the original proof (J. SPENCER [1972*b*]) for details. We assume $n = 2m + 1$, the case n even being similar. We leave the lower bound as Exercise 1.

If $T = T_n$ is a tournament and $i \in [n]$ set $W_i = \{j : (i, j) \in T\}$, $L_i = \{j : (j, i) \in T\}$, $s_i = |W_i| =$ the score of i. The ordered n-tuple $(s_1, \ldots, s_n)$ is called the score sequence of T. If all $s_i = m$, T is called regular. Set $\mathfrak{R} =$ the class of regular tournaments on $[n]$. Define $\mathbf{T}^r$ as a random member of $\mathfrak{R}$.

We want to get a handle on the properties of $\mathbf{T}^r$. For example, we know Prob $\big((1, 2), (2, 3), (3, 1) \in \mathbf{T}\big) = 1/8$. What is the probability when $\mathbf{T}$ is replaced by $\mathbf{T}^r$? Intuitively, we feel that the probability should be greater than 1/8 but only by a very slight amount. Our starting point is

THEOREM 10.2: *Let $f(s_1, \ldots, s_n)$ be the number of tournaments with score sequence $(s_1, \ldots, s_n)$. If $|s_i - m| < n^{3/5}$ for $1 \leq i \leq n$ then*

$$f(s_1, \ldots, s_n) =$$
$$= f(s_1 + 1, s_2 - 1, \ldots, s_n)\left(1 + \frac{(s_1 - s_2 + 1)}{n}(4 + o(1))\right).$$

The proof (J. SPENCER [1972]), is quite involved and will not be given. Since f is a symmetric function the equation also applies when s_i, s_j are replaced by $s_i + 1$, $s_j - 1$. The condition $|s_i - m| < n^{3/5}$ could be made weaker, but this would not affect our results. If $\sum s_i = \binom{n}{2}$ and $|s_i - m| < n^{3/5}$ then repeated application of Theorem

10.2 yields

$$f(s_1, \ldots, s_m) = f(m, \ldots, m)e^{-(2+o(1))\Sigma(s_i-m)^2/n}. \tag{10.1}$$

For $A \subseteq [n]$ define

$$T_A = T \cap \{(x, y) : x \in A \text{ or } y \in A\},$$
$$T \mid A = T \cap \{(x, y) : x \in A \text{ and } y \in A\}.$$

Set

$$\mathcal{S}(A) = \{T \mid A : T \in \mathcal{R}\},\ \mathcal{G}(A) = \{T_A : T \in \mathcal{R}\}.$$

We shall restrict $|A| = k < n^{3/5}$. Then $\mathcal{S}(A) =$ the class of tournaments on A and $\mathcal{G}(A)$ is the class of sets of games played by A given each $i \in A$ winning exactly m games.

COROLLARY 10.1: *Let* $|A| = k \leq n^{3/5}$, $T_1 \in \mathcal{G}(A)$. *Then*

$$|\mathcal{G}(A)|^{-1}e^{-k^2/2} \leq \text{Prob}\,[\mathbf{T}^r_A = T_1] \leq |\mathcal{G}(A)|^{-1}e^{k^2/2}.$$

PROOF: For notational convenience suppose $A = \{1, \ldots, k\}$. Given T_1 we extended to a $T \in \mathcal{R}$ by adding a subtournament T_2 on $[n] - A$. $T \in \mathcal{R}$ iff each $i \in A^c$ wins exactly $x_i = m - |\{a \in A : (i, a) \in T_1\}|$ games in B. So $\left|x_i - \dfrac{n-k}{2}\right| < \dfrac{k}{2}$ and

$$|\{T \in \mathcal{R} : T_A = T_1\}| = f(x_{k+1}, \ldots, x_n). \tag{10.2}$$

By (10.1) the maximum ratio of two such $f(x_{k+1}, \ldots, x_n)$ is $e^{k^2/2}$, and Corollary 10.1 follows.

COROLLARY 10.2: *Let* $|A| = k \leq n^{3/5}$, $T_1 \in \mathcal{S}(A)$. *Then*

$$|\mathcal{S}(A)|^{-1}e^{-2k^3/n} \leq \text{Prob}\,[\mathbf{T}^r \mid A = T_1] \leq |\mathcal{S}(A)|^{-1}e^{2k^3/n}.$$

PROOF: As in Corollary 10.1, it suffices to show that the maximum ratio of Prob $[\mathbf{T}^r \mid A = T_1]$ for two values of T_1 is bounded by $e^{2k^3/n}$. If $i, j \in A$, set $T_1^{ij} = T_1 \Delta \{(i, j), (j, i)\}$. Since any T_2 can be reached from T_1 by $\binom{k}{2}$ such shifts it suffices to show

$$\text{Prob}\,[\mathbf{T}^r \mid A = T_1] \geq \text{Prob}\,[\mathbf{T}^r \mid A = T_1^{ij}]e^{-4k/n}. \tag{10.3}$$

That is,

$$|\{T \in \mathcal{R} : T \mid A = T_1\}| \geq |\{T \in \mathcal{R} : T \mid A = T_1^{ij}\}|\, e^{-4k/n}. \tag{10.4}$$

Set $B = A - \{i, j\}$. It suffices to show for every value of $T'_B \in \mathcal{G}(B)$ consistent with $T \mid A = T_1$ that, setting

$$\begin{aligned} H(T_1, T'_B) &= \{T \in \mathcal{R} : T \mid A = T_1, T_B = T'_B\} \\ H(T_1^{ij}, T'_B) &= \{T \in \mathcal{R} : T \mid A = T_1^{ij}, T_B = T'_B\}, \end{aligned} \tag{10.5}$$

we have

$$| H(T_1, T'_B) | \geq | H(T_1^{ij}, T'_B) | e^{-4k/n}. \tag{10.6}$$

We may assume $(i, j) \in T_1$. Let $i = 1$, $j = 2$, $A = [k]$ for convenience. Let $m - a$, $m - b$ = the scores of i, j in T_1. Then in T_1^{ij} they have scores $m - a - 1$, $m - b + 1$. For $t \notin A$ set

$$x_t = m - 1 - | \{y \in B : (t, y) \in T'_B\} | . \tag{10.7}$$

Then

$$\begin{aligned} | H(T_1, T'_B) | &= \sum_{(K,t)} f(x_{k+1} - \varepsilon_{k+1}, \ldots, x_n - \varepsilon_n) \binom{t}{t/2 + \delta} \\ | H(T_1^{ij}, T'_B) | &= \sum_{(K,t)} f(x_{k+1} - \varepsilon_{k+1}, \ldots, x_n - \varepsilon_n) \binom{t}{t/2 + \delta + 1} \end{aligned} \tag{10.8}$$

where $K = a + b + k - n$, $\delta = (a - b)/2$, $\sum_{(K,t)}$ is the summation over all $\varepsilon_{k+1}, \ldots, \varepsilon_n = 1, 0$, or -1 with sum K and exactly t zeroes. The proof of these statements, whose details we omit, consists of first showing that T_{A^c} must have score sequence of the form $x_{k+1} - \varepsilon_{k+1}, \ldots, x_n - \varepsilon_n$ with $\sum \varepsilon_i = K$ and that, given this score sequence, T may be completed in $\binom{t}{t/2 + \delta}$ or $\binom{t}{t/2 + \delta + 1}$ ways respectively. By elementary calculations Theorem 10.2, (10.8), and $| \delta | \leq k$ imply (10.5) and therefore Corollary 10.2.

Corollaries 10.1, 10.2 are used to show that $\mathbf{T}^r$ has many of the "local" properties that $\mathbf{T}$ has. We say T has property I if there do not exist disjoint A, $B \subseteq [n]$, $| A | = | B | = t = [3 \log_2 n]$, $A \times B \subseteq T$.

LEMMA 10.1: Prob [$\mathbf{T}^r$ has property I] $> 1 - o(1)$.

PROOF:

$$\begin{aligned} &\text{Prob}\,[\mathbf{T}^r \text{ does not have property I}] \\ &< \sum_{A,B} \text{Prob}\,[A \times B \subseteq \mathbf{T}^r], \end{aligned} \tag{10.9}$$

the summation over $| A | = | B | = t$, $A \cap B = \emptyset$. Fix such A, B

$$\begin{aligned} \text{Prob}\,[A \times B \subseteq \mathbf{T}^r] &= \sum_{T_1 \supseteq A \times B} \text{Prob}\,[\mathbf{T}^r \mid (A \cup B) = T_1] \\ &\leq 2^{\binom{2t}{2} - t^2} \max_{T_1} \text{Prob}\,[\mathbf{T}^r \mid A \cup B = T_1] \\ &\leq 2^{\binom{2t}{2} - t^2}\, 2^{-\binom{2t}{2}} e^{2(2t)^2/n} \\ &\leq 2^{-t^2} (1 + o(1)) . \end{aligned} \tag{10.10}$$

The crucial step is using Corollary 10.2 to show

$$\text{Prob}\,[\mathbf{T}^r \mid A \cup B = T_1] \leq \text{Prob}\,[\mathbf{T} \mid A \cup B = T_1]\big(1 + o(1)\big).$$

The rest of the proof proceeds as with $\mathbf{T}$. For

$$(10.11) \qquad \sum_{A,B} \text{Prob}\,[A \times B \subseteq \mathbf{T}^r] \leq n^{2t}\, 2^{-t^2}\big(1 + o(1)\big) = o(1)\,.$$

We need a similar lemma. Say T has property II if A, B disjoint, $|A| = |B| = t \geq 10 \log n$ implies $|(A \times B) \cap T| \geq 40t$.

LEMMA 2: Prob [$\mathbf{T}^r$ has property II] $= 1 - o(1)$.

PROOF: Again we show

$$(10.12) \qquad \text{Prob}\,[\mathbf{T}^r \text{ does not have property II}] = o(1)\,.$$

Suppose $|A| = |B| = t \geq 10 \log n$ and $|(A \times B) \cap T| < 40t$. Set $t_0 = [10 \log n]$. By Exercise 2 there exists disjoint

$$A_0, B_0, |A_0| = |B_0| = t_0, |(A_0 \times B_0) \cap T| < 40t_0\,.$$

Thus it suffices to show

$$(10.13) \quad \text{Prob}\,[\exists A_0, B_0 \in [n]^{t_0}, A_0 \cap B_0 = \varnothing, |(A_0 \times B_0) \cap \mathbf{T}^r| < 40t_0] = o(1)$$

which reduces to

$$(10.14) \qquad \text{Prob}\,[|(A_0 \times B_0) \cap \mathbf{T}^r| < 40t_0] = o(n^{-2t_0})$$

which follows as in Lemma 10.1.

We need Corollary 10.1 for a more complex lemma. Let $S \subseteq [n]$, $i, j \in [n]$, $i \neq j$. For fixed T write (Sij) if either

(1) $i, j \in S$ and $|(S^c \cap W_i) - (S^c \cap W_j)| < 200 \log n$

or

(2) $i, j \in S^c$ and $|(S \cap W_i) - (S \cap W_j)| < 200 \log \text{n}$.

Say T has property III if $||S| - n/2| < 5\sqrt{n}\sqrt{\log n}$ implies there exist $i_1, \ldots, i_{10} \in [n]$ such that (Sjk) implies $\{j, k\} \cap \{i_1, \ldots, i_{10}\} = \varnothing$.

LEMMA 10.3: Prob [$\mathbf{T}^r$ has property III] $= 1 - o(1)$.

PROOF: If $\mathbf{T}^r$ does not have property III it must have property $Q =$ "$\exists$ distinct $i_1, \ldots, i_5, j_1, \ldots, j_5$ $(Si_t j_t)$ for $1 \leq t \leq 5$". Therefore it suffices to show

$$(10.15) \qquad \text{Prob}\,[\mathbf{T}^r \text{ has property } Q] = o(1)\,.$$

As there are $< 2^n n^{10}$ choices of $S, i_1, \ldots, j_5$ it suffices to show for fixed $S, i_1, \ldots, j_5$

$$\text{Prob}\,[(Si_t j_t) \text{ for } 1 \leq t \leq 5] = o(2^{-n} n^{-10})\,. \tag{10.16}$$

From Corollary 10.1 we can assume, within a $1 + o(1)$ factor that the W_{i_t}, W_{j_t} are random subsets of m elements. Thus

$$\text{Prob}\,[(Sij)] < \left(\frac{3}{4} + o(1)\right)^n \tag{10.17}$$

so

$$\text{Prob}\,[(Si_t j_t) \text{ for } 1 \leq t \leq 5] < \left(\frac{3}{4} + o(1)\right)^{5n} = o(.487^n) = = o(2^{-n} n^{-10})\,. \tag{10.18}$$

From Lemmas 10.1, 10.2, 10.3 we find, and fix, T such that

$$\begin{gathered} T \in \mathfrak{R} \\ T \text{ has properties I, II, III.} \end{gathered} \tag{10.19}$$

The remainder of the proof is nonprobabilistic. Let P be a partial order, $P \subseteq \text{T}$. We need to show

$$|\,P\,| \leq \frac{n^2}{8} + c_1 \log n\,. \tag{10.20}$$

We show that P is "basically" a subset of a bipartite graph. Let

$$\begin{gathered} L = \{x : |\,\{y : (y, x) \in P\}\,| > 10 \log n\} \\ W = L^c. \end{gathered} \tag{10.21}$$

All players are winners (W) or losers (L). The winners lose less than $10 \log n$ games in P by definition. Let $x \in L$, $A = \{z : (x, z) \in P\}$, $B = \{y : (y, x) \in P\}$. Then $B \times A \subseteq P$ so, since $|\,B\,| > 10 \log n$, $|\,A\,| < 10 \log n$ by I. For any A

$$T \cap (A \times A^c) = \sum_{i \in A} s_i - \binom{|\,A\,|}{2} = \frac{1}{2}\,|\,A\,|\;|\,A^c\,| \tag{10.22}$$

since all $s_i = m$. Thus, setting $|\,W\,| = \frac{n}{2} + \alpha$, $|\,L\,| = \frac{n}{2} - \alpha$,

$$|\,P \cap (W \times L)\,| \leq \frac{1}{2}\,|\,W\,|\;|\,L\,| \leq \frac{n^2}{8} - \frac{\alpha^2}{2}\,. \tag{10.23}$$

(This is where we need $T \in \mathfrak{R}$.) If $x \in L$, $z \in W$, and $(x, z) \in P$ then $(y, z) \in P$ for the at least $10 \log n$ y's such that $(y, x) \in P$, a contradiction. Thus $|\,P \cap (L \times W)\,| = 0$. $|\,P \cap (W \times W)\,| = \sum_{z \ni W} |\{y \in$

$W : (y, z) \in P\} \mid \leq (10 \log n) \mid W \mid$. Similarly $\mid P \cap (L \times L) \mid \leq (10 \log n) \mid L \mid$ so

$$\mid P \mid \leq \frac{n^2}{8} - \frac{\alpha^2}{2} + 10\, n \log n\,. \tag{10.24}$$

We assume $\alpha \leq \sqrt{20\, n \log n}$ since otherwise (10.20) follows trivially. By property III we find, and fix $S = \{i_1, \ldots, i_{10}\}$ so that (Wxy) implies $\{x, y\} \cap \{i_1, \ldots, i_{10}\} = \emptyset$.

Set $P' = (W \times L) \cap T$. We compare $\mid P \mid$ and $\mid P' \mid$. The elements of $P - P'$ are of the form (x, y), $x, y \in W - S$ or (s, x), $s \in W \cap S$, $x \in W - S$, or the analogous pairs for L, plus at most $100 \log n$ pairs of the form (x, s), $s \in W \cap S$, $x \in W - S$ and (s, x), $s \in L \cap S$, $x \in L - S$. For each $y \in W - S$ there are at most $10 \log n$ $x \in W - S$, $(x, y) \in P$. If there is at least one such x then for all $z \in (L - S) \cap (W_y - W_x)$ we have $(y, z) \in T$, $(x, z) \notin T$ so $(y, z) \in P - P'$. By property III there are at least $200 \log n$ such (y, z). If $s \in S \cap W$ such that $\mid \{x \in W : (s, x) \in P\} \mid < 20 \log n + 2\alpha$ we ignore these $(s, x) \in P - P'$, (there are at most $200 \log n + 20\alpha$ such (s, x)). If not, set $B_s = \{x \in W : (s, x) \in P\}$, $C_s = \{y \in L : (y, s) \in T\}$ and $H = (B_s \times C_s) \cap T$. By definition, $H \subseteq P'$. If $(x, y) \in P$, $(s, x) \in P$ implies $(s, y) \in P \subseteq T$, a contradiction. Thus $H \subseteq P' - P$. As $B_s' \cup (L - C_s) \subseteq W_s$ and $\mid W_s \mid = m$, $\mid L \mid = \frac{n}{2} - \alpha$ we have $\mid C_s \mid \geq \mid B_s \mid - \alpha \geq 20 \log n$. By property II, $\mid H \mid \geq 20 \mid B_s \mid$.

For each pair in $P - P'$, with at most $300 \log n + 20\alpha$ exceptions, we have associated at least 20 pairs in $P' - P$. There has been some multiple counting but it can be verified that no pair in $P' - P$ was counted more than 12 times. Thus

$$\mid P \mid - \mid P' \mid \leq 300 \log n + 20\alpha. \tag{10.25}$$

By (10.24),

$$\mid P' \mid \leq \frac{1}{2} \mid W \mid \; \mid L \mid \leq \frac{n^2}{8} - \frac{\alpha^2}{2}. \tag{10.26}$$

Therefore

$$\begin{aligned} \mid P \mid &\leq \frac{n^2}{8} + \left(20\alpha - \frac{\alpha^2}{2}\right) + 300 \log n \\ &\leq \frac{n^2}{8} + 301 \log n\,. \end{aligned} \tag{10.27}$$

Q.E.D.

The essential idea in this proof is to manipulate $\mathbf{T}^r$. With Corollaries 10.1 and 10.2 we are able to handle $\mathbf{T}^r$ similarly to $\mathbf{T}$. A similar random variable appears in Sec. 14, Asymmetric Graphs.

Exercises

1. Let T be a fixed tournament on $[n]$. Show that there exists partial order $P \subseteq T$, $|P| \geq \frac{n^2}{8} + \left(\frac{1}{2} - o(1)\right) \log_2 n$. (Hint: If no $s_i = m$ set $W =$ the m players with highest scores, $P = (W \times W^c) \cap T$. Suppose $s_i = m$. If $|(L_i \times W_i) \cap T| < n^2\left(\frac{1}{8} - o(1)\right)$ set $P = (W_i \times \times L_i) \cap T$. Otherwise, by Corollary 12.1, find $A \subseteq L_i$, $B \subseteq W_i$, $|A| = |B| \geq (1 - o(1)) \log_2 n$, $B \times A \subseteq T$. Setting $W = W_x + A - B$ either $P_1 = (W \times W^c) \cap T$ or $P_2 = ((W^c \times W) \cap T) \cap \{(b, i) : b \in B\}$ is sufficiently large.)

2. Let $A \cap B = \emptyset$, $G \subseteq A \times B$, $|A| = a$, $|B| = b$. Fix $0 \leq c \leq a$, $0 \leq d \leq b$. Prove there exist $C \in [A]^c$, $D \in [B]^d$

$$|G \cap (C \times D)| \geq |G| (c/a) (d/b) .$$

11. CHROMATIC NUMBER

In this section we consider bounds on the chromatic number $\chi(G)$. We will use G that are values of a random variable $\mathbf{G}$. We use two methods to get bounds on $\chi(G)$.

Method 1: If a graph on n points can be colored in k colors then some color is used at least n/k times. Therefore the graph contains a set of at least n/k independent points. Using this method we can show that a graph with relatively few edges has high chromatic number.

Method 2: Suppose $\chi(G) \geq k$. Let S be a minimal set such that $\chi(G \mid S) = k$. For all $x \in S$, $\chi(G \mid S - \{x\}) \leq k - 1$. If x had degree $< k - 1$ in S then the $(k - 1)$ coloring of $S - \{x\}$ could be extended to a $(k - 1)$ coloring of S, a contradiction. Thus each point $x \in S$ has degree $\geq k - 1$ in S. Consequently, if $\chi(G) \geq k$ there exists u and a subgraph on u points with $\geq u(k - 1)/2$ edges. This method yields an upper bound on the chromatic number.

We now consider a problem whose solution requires both methods. Define $f(m, k, n)$ as the maximal $\chi(G)$ of a graph G on n points such that all m point subgraphs may be k-colored. We are interested in the case $k = 3$.

THEOREM 11.1 (P. ERDŐS [1962]): *There exists $C_1 > 0$ such that, for $n \geq m \geq 3$*

$$f(m, 3, n) \geq C_1(n/m)^{1/3} \left(\log(2n/m)\right)^{-1}.$$

In particular, this implies the perhaps surprising result that for every k there exists $\varepsilon > 0$ and a graph G on n points $\left(n > n(k, \varepsilon)\right)$ such that $\chi(G) \geq k$, yet every subgraph of $[\varepsilon n]$ vertices may be 3-colored.

First, we dispense with the "small" cases. Set

$$C_2 = \min_{\substack{3 \leq n \leq 10^{10} \\ 3 \leq m \leq n}} f(m, 3, n)/\left[(n/m)^{1/3} \left(\log(2n/m)^{-1}\right)\right]. \tag{11.1}$$

Then $C_2 > 0$. If $C_1 \leq C_2$ Theorem 11.1 holds for $3 \leq n \leq 10^{10}$. Similarly, set

$$C_3 = \left[\sup_{1 \leq \alpha \leq 10^{10}} \alpha^{1/3} \left(\log(2\alpha)\right)^{-1}\right]^{-1}. \tag{11.2}$$

Here α represents n/m. If $C_1 \leq C_3$ then

$$f(m, 3, n) \geq 1 \geq C_1(n/m)^{1/3}\left(\log(2n/m)\right)^{-1}$$

for $n/m \leq 10^{10}$. This is the only point in the proof where $\log(2n/m)$ could not be replaced by $\log(n/m)$. Thus, to prove Theorem 11.1 it suffices to find $C_4 > 0$ so that $n > 10^{10}$, $n/m > 10^{10}$ imply

$$f(m, 3, n) > C_4(n/m)^{1/3}\left(\log(2n/m)\right)^{-1}$$

and then set $C_1 = \min(C_2, C_3, C_4)$.

Fix $n > 10^{10}$, $m < n10^{-10}$. Let $\mathbf{G} = \mathbf{G}_{n,p}$ with p as yet undetermined. We apply Method 2. If a subgraph of $\mathbf{G}$ on m points required 4 colors then some subgraph on u points, $4 \leq u \leq m$, would contain $\geq 3u/2$ edges. $\mathbf{G}$ contains such a subgraph with probability

$$\begin{aligned} g_p(n, m) &\leq \sum_{u=4}^{m} \sum_{A \in [n]^u} \text{Prob}\,[|\,\mathbf{G} \mid A\,| \geq 3u/2] \\ &\leq \sum_{u=4}^{m} \binom{n}{u} \binom{\binom{u}{2}}{\{3u/2\}} p^{\{3u/2\}} \\ &\leq \sum_{u=4}^{m} \left(\frac{ne}{u}\right)^{u} \left(\frac{u^2}{3u}\right)^{3u/2} p^{3u/2} \\ &\leq \sum_{u=4}^{m} [C_5 n u^{1/2} p^{3/2}]^u \\ &\leq \sum_{u=4}^{m} [C_5 n m^{1/2} p^{3/2}]^u \\ &< 1/10 \end{aligned} \tag{11.3}$$

when $C_5 n m^{1/2} p^{3/2} < 1/10$, which holds if $p \leq C_6 n^{-2/3} m^{-1/3}$. Now fix $p = C_6 n^{-2/3} m^{-1/3}$.

$$\text{Prob}\,[\text{if } |\,A\,| \leq m,\ \chi(\mathbf{G} \mid A) \leq 3] > 9/10 \tag{11.4}$$

To bound $\chi(\mathbf{G})$ we use Method 1. The probability that $\mathbf{G}$ contains a set of x independent points is

$$\begin{aligned} h_p(n, m) &< \sum_{A \in [n]^x} \text{Prob}\,[\mathbf{G} \cap [A]^2 = \varnothing] \leq \binom{n}{x} (1 - p)^{\binom{x}{2}} \\ &< [(ne/x)(1 - p)^{(x-1)/2}]^x. \end{aligned} \tag{11.5}$$

Since $n > 10^{10}$ ("sufficiently large"), $p < C_6 n^{-2/3} < 10^{-5}$ ("sufficiently small"). As $n/m > 10^{10}$, $pn = C_6(n/m)^{1/3} > 10^2$. We can now show, with "room to spare", that for $x \geq 4 \log(pn)/p$, $h_p(n, m) < 1$.

If G does not contain x independent points, $\chi(\mathbf{G}) \geq n/x$. Thus

$$\text{Prob}\,[\chi(\mathbf{G}) \geq pn/4 \log (pn)] > 9/10 \tag{11.6}$$

Combining (11.4) and (11.6) we find G such that all subgraphs of $\leq m$ points can be 3-colored and

$$\chi(G) \geq pn/4 \log(pn) = \frac{3}{4} C_6(n/m)^{1/3}[\log(n/m)]^{-1}$$
$$> C_4(n/m)^{1/3}(\log(2n/m))^{-1}$$

for $C_4 = \frac{3}{4} C_6$. Q.E.D.

We now look at a theorem that requires a refinement of Method 1. Set $\omega(G)$ = the clique number of G. In any coloration of G each element of a clique must be colored distinctly. Thus, $\chi(G) \geq \omega(G)$. W. Tutte [1947], A. A. Zykov [1949], J. B. Kelly and L. B. Kelly [1954] independently constructed, for each m, a graph G with $\omega(G) = 2$, $\chi(G) \geq m$. In fact, the G constructed had no circuits of length ≤ 5. J. B. Kelly and L. B. Kelly [1954] conjectured that for every k, l there exists a graph with $\chi(G) \geq k$ and no circuits of length $\leq l$. This was first shown by P. Erdős using a probabilistic proof. A constructive proof was later found by L. Lovász [1968]. Erdős' proof gave the following result which is much stronger.

Theorem 11.2 (P. Erdős [1959]): *Fix l, θ, η, $\theta < 1/l$, $0 < \eta < \theta/2$. For n sufficiently large there exists a graph G on n points with no circuits of length $\leq l$ and $\chi(G) \geq n^{\eta}$.*

Proof: Let $\mathbf{G} = \mathbf{G}_{n,p}$, $p = n^{\theta-1}$. Set $x = [n^{1-\eta}]$. Set

$$\alpha = \text{Prob}\,[A \in [n]^x \text{ implies } |\, \mathbf{G} \,|\, A \,| \geq n]\,.$$

Then

$$1 - \alpha \leq \sum_{A \in [n]^x} \text{Prob}\,[|\, \mathbf{G} \,|\, A \,| < n]$$

(11.7)

$$\leq \binom{n}{x} \binom{\binom{x}{2}}{n} (1-p)^{\binom{x}{2}}.$$

We bound

$$\binom{n}{x} \leq n^x \leq n^n, \binom{\binom{x}{2}}{n} < \left(\frac{ex^2}{n}\right)^n < n^n, \tag{11.8}$$

and approximate

$$(1-p)^{\binom{x}{2}} \sim e^{-px^2/2} \sim e^{-\frac{1}{2}n^{1+(\theta-2\eta)}} \tag{11.9}$$

Since $\theta - 2\eta > 0$ and independent of n, $1 - \alpha = o(1)$. Therefore, $\alpha = 1 - o(1)$.

Now we examine the small circuits of $\mathbf{G}$. An i-circuit is determined by an i-tuple $(x_1, \ldots, x_i)$ of distinct vertices of $[n]$. There are $< n^i$ of these i-tuples each of which gives a circuit iff $(x_j, x_{j+1}) \in \mathbf{G}$, $1 \leq \leq j \leq i \pmod{i}$, which occurs with probability p^i. Set $c(G) =$ the number of edges of G contained in i-circuits, $3 \leq i \leq l$. Then

$$\mathsf{E}\big(c(\mathbf{G})\big) \leq \sum_{i=3}^{l} i n^i p^i \leq l^2 n^{\theta l} = o(n) . \tag{11.10}$$

So with probability $1 - o(1)$, $c(\mathbf{G}) < n$.

Combining the above results we find a value G' of $\mathbf{G}$ such that

(1) $A \in [n]^x$ implies $| G' \mid A | \geq n$

(2) $c(G') < n$.

Set $G = G'$ — all edges in circuits of length $\leq l$. By definition, G has no circuits of length $\leq l$. As $c(G') < n$, fewer than n edges have been deleted from G and consequently $A \in [n]^x$ implies $G \mid A \neq \varnothing$. By Method 1 this implies $\chi(G) \geq n/x \geq n^\eta$. Q.E.D.

We turn to a comparison of $\omega(G_n)$ and $\chi(G_n)$. We first examine max $\chi(G_n)$ over those G_n with $\omega(G_n) \leq 2$. From Theorem 5.3 there exists a G_n with

$$\omega(G_n) = 2, \ \omega(G_n^c) \leq C_7 n^{1/2} \log n .$$

Thus

$$\chi(G_n) \geq n/\omega(G_n) \geq C_8 n^{1/2}/\log n .$$

The lower bound on $\chi(G_n)$ requires the following lemma.

Lemma 11.1: *Let G_n be fixed. Suppose that for each $A \in [n]^x$, there exists $B \subseteq A$, $| B | \geq f(x)$, $G_n \mid B = \varnothing$. ($1 \leq x \leq n$, x, $f(x)$ integral.) Set $a_0 = n$, $a_{i+1} = a_i - f(a_i)$. Then if $a_t = 0$, $\chi(G_n) \leq t$.*

Proof: By induction we find $A_i \subseteq [n] - A_1 - \ldots - A_{i-1}$, $| A_i | = f(a_{i-1})$, $G \mid A_i = \varnothing$. Then $[n] = A_1 + \ldots + A_t$ is a t-coloring of G_n. Q.E.D.

Suppose $\omega(G_n) \leq 2$. Then any subgraph G_m has $\omega(G_m) \leq 2$. Since from Theorem 5.3 $R(C_9[m \log m/\log\log m]^{1/2}, 3) \leq m$, G_m containe a set of $C_9[m \log m/\log \log m]^{1/2}$ independent points. Applying Lemma 11.1, $\chi(G) \leq C_{10}[n \log\log n/\log n]^{1/2}$. We summarize by

Theorem 11.3: Set $\mathcal{G}_n = \{G_n : \omega(G_n) \leq 2\}$.

$$C_8 n^{1/2}/\log n \leq \max_{G \in \mathcal{G}_n} \chi(G) \leq C_{10} n^{1/2} [\log \log n/\log n]^{1/2}.$$

We can also look at max $\chi(G_n)/\omega(G_n)$. By Theorem 5.1 there exists G_n, with $\omega(G_n)$ and $\omega(G_n^c) \leq \log_4 n$. Since $\chi(G_n) \geq n/\omega(G_n^c)$, $\chi(G_n)/\omega(G_n) \geq \geq C_{11} n/(\log n)^2$. By (5.2), for all G_m

$$\binom{\omega(G_m) + \omega(G_m^c) - 2}{\omega(G_m) - 1} \geq m . \tag{11.11}$$

LEMMA 11.2: *If* $\binom{a+b}{b} \geq m$, $ab \geq C_{12}(\log m)^2$.

PROOF: By symmetry we may assume $\alpha = a/b \geq 1$.

$$m \leq \binom{a+b}{b} \leq \binom{2a}{b} \leq (2a)^b/(b/e)^b = (2e\alpha)^b$$

$$ab = b^2\alpha \geq (\log m)^2\alpha/(\log 2e\alpha)^2 \geq C_{12}(\log m)^2$$

where $C_{12} = \min_{\alpha \geq 1} \alpha/(\log 2e\alpha)^2 = e/8$. Q.E.D.

Applying Lemma 11.2 to (11.11) with $a = \omega(G_m) - 1$, $b = \omega(G_m^c) - 1$

(11.12) $$\omega(G_m)\omega(G_m^c) \geq ab \geq C_{12}(\log m)^2 .$$

Now let G_n be fixed. Set $\omega = \omega(G_n)$. The conditions of Lemma 11.1 hold with $f(m) = \{C_{12}(\log m)^2/\omega\}$. By elementary methods, Lemma 11.1 implies

(11.13) $$\chi(G_n) \leq C_{13} n\omega(G_n)/(\log n)^2 .$$

Again summarizing:

THEOREM 11.4 (P. ERDŐS [1967]):

$$C_{11}\frac{n}{(\log n)^2} \leq \max \chi(G_n)/\omega(G_n) \leq C_{13}\frac{n}{(\log n)^2} .$$

Exercises

1. Fix $l \geq 4$. Set

$$f_l(n) = \max_{G \in \mathcal{G}} \chi(G)$$

where $\mathcal{G}$ is the class of graphs on n points with no cycles of length $\leq l$. By Theorem 11.2, if $\eta < 1/2l$ then $f_l(n) \geq n^\eta$ for n sufficiently large. Tighten the proof of Theorem 11.2 to raise the lower bound for $f_l(n)$.

2. Using the notation of the previous exercise, show

$$f_l(n) \leq n^{1/[l/2]} + 2 .$$

(Hint: Count points on the subgraph given by Method 2.)

3. Let **G** be a random graph on n points with edge probability $= 1/2$. Prove that there exist positive constants c_1, c_2 such that with probability $1 - o(1)$

$$\frac{c_1 n}{\log n} < \chi(\mathbf{G}) < \frac{c_2 n}{\log n} .$$

4. (P. Erdős and A. Hajnal [1966]) (see § 4 for notation). An s-circuit in a k-graph $\mathfrak{J}$ is defined as a family $\{A_1, \ldots, A_s\}$ of edges of $\mathfrak{J}$ such that $\left| \bigcup_{i=1}^{s} A_i \right| \leq (k-1)\,s$. Fix s, k. Prove that there exists a k-graph $\mathfrak{J}$ with no t-circuits, $t \leq s$, such that $\mathfrak{J}$ does not have property B. (Hint: This is a generalization of Theorem 11.2. Let n be large and choose $\mathbf{I} \subseteq [n]^k$ by placing each $A \in \mathbf{I}$ with probability $p = n^{1-k+\delta}$, δ very small. Then the expected number of $\leq s$-circuits is $< Cn^{s\delta}$. With probability $1 - o(1)$ every $n^{1-\varepsilon}$ point contains $\sim n^{1+\eta-\varepsilon k}$ members of $\mathfrak{J}$. Delete small circuits from $\mathfrak{J}$ to find $\mathfrak{J}'$ with $\chi(\mathfrak{J}') \geq$ $\geq n^{\varepsilon} > 2$ for $\varepsilon < \big(1 - (s-1)\,\eta\big)/p$.)

5. Note that Exercise 4, with $s = 2$, implies Exercise 3 of § 4. Define $m^*(n)$ as the minimal $|\mathfrak{J}|$ of an n-graph $\mathfrak{J}$ not having property B with $A_1, A_2 \in \mathfrak{J}$ implying $|A_1 \cap A_2| \leq 1$. Prove $m^*(n) \leq K^n$, where K is an absolute constant.

6. (?) Does $\lim\limits_{n} \dfrac{\log m^*(n)}{n} = \log 2$?

7. Let G be a tripartite graph on disjoint sets A, B, C of cardinality n. That is, $G \subseteq [A \cup B \cup C]^2, G \mid A = G \mid B = G \mid C = \emptyset$. N. Sauer (private correspondence) asked whether G must contain either a triangle or $n + 1$ independent points.

(a) Prove that for n sufficiently large there exist tripartite G containing neither triangles nor $n + 1$ independent points. [Hint: Set $\mathbf{G} = \mathbf{G}_p - [A]^2 - [B]^2 - [C]^2$, $p = n^{\varepsilon-1}$ and $\mathbf{G}^* = \mathbf{G} -$ all triangles.]

(b) Let k be a fixed integer. Prove that for n sufficiently large there exist tripartite G containing no cycles of length $\leq k$ nor $n + 1$ independent points.

8. (?) The following question of B. Grunbaum is unsolved: Is it true that for every k and l there is a regular graph of valency k, chromatic number k, whose smallest circuit has $\geq l$ edges? This seems very difficult and probably cannot be done by probabilistic methods.

12. ZARANKIEWICZ'S PROBLEM AND BIPARTITE RAMSEY THEOREMS

Definition: Assume $2 \leq a \leq m$, $2 \leq b \leq n$, a, b, m, n integral. Then $k_{ab}(m, n)$ is the minimal k such that all $m \times n$ $0-1$ matrices containing at least k ones contain an $a \times b$ submatrix consisting entirely of ones. Set $k_a(m, n) = k_{aa}(m, n)$, $k_{ab}(n) = k_{ab}(n, n)$, $k_a(n) = k_{aa}(n, n)$ for convenience.

ZARANKIEWICZ's [1951] Problem is to find $k_{ab}(m, n)$. R. GUY and S. ZNÁM [1969] give applications and connections of ZARANKIEWICZ's Problem to affine and projective planes, Steiner-triples, and other areas. R. GUY and S. ZNÁM [1971] give a thorough account of the known bounds on the function k. We are particularly interested in a graph theoretic interpretation. Fix disjoint sets A_1, A_2, $|A_1| = m$, $|A_2| = n$.

An $m \times n$ $0-1$ matrix may be thought of as the adjacency matrix of a bipartite graph $G \subseteq A_1 \times A_2$. (Throughout this section we think of the elements of $A_1 \times A_2$ as unordered pairs.) Then $k = k_{ab}(m, n)$ is the minimal integer such that if $G \subseteq A_1 \times A_2$, $|G| \geq k$, there exist $B_i \subseteq A_i$, $|B_1| = a$, $|B_2| = b$, $B_1 \times B_2 \subseteq G$.

To find bounds on $k_{ab}(m, n)$ it will be convenient to consider a more general function.

Definition: $B(m, n, a, b; e)$ is the largest integer such that if $|A_1| = m$, $|A_2| = n$, $A_1 \cap A_2 = \varnothing$, $G \subseteq A_1 \times A_2$, $|G| \geq e$, there are at least $B(m, n, a, b; e)$ pairs (B_1, B_2), $B_1 \in [A_1]^a$, $B_2 \in [A_2]^b$, $B_1 \times B_2 \subseteq G$.

Clearly,

$$B(m, n, a, b; e) > 0 \text{ iff } k_{ab}(m, n) \geq e . \tag{12.1}$$

THEOREM 12.1:

$$\binom{m}{a}\binom{n}{b}\frac{\binom{mn-ab}{e-ab}}{\binom{mn}{e}} \geq B(m, n, a, b; e) \tag{12.2}$$

$$\geq \binom{n}{b}\binom{m\binom{e/m}{b}\Big/\binom{n}{b}}{a}$$

The upper bound is the expected number of $B_1 \times B_2 \subseteq \mathbf{G}$ where $\mathbf{G}$ is a random member of $[A_1 \times A_2]^e$. (See Exercise 1.) For the lower bound we require some elementary results on convex functions. Let t be a nonnegative integer. The function $\binom{x}{t}$ is defined for x real in § 2.

Lemma 12.1: *If $f(x)$ is convex, then $g(x) = \binom{f(x)}{t}$ is convex.*

Proof: By calculating the second derivative, it is easily shown that $f_1(x) = \binom{x}{t}$ is convex. If $0 \leq \lambda \leq 1$

$$\lambda \binom{f(x)}{t} + (1-\lambda) \binom{f(y)}{t} \geq \binom{\lambda f(x) + (1-\lambda) f(y)}{t}$$

(as f_1 is convex)

$$\geq \binom{f(\lambda x + (1-\lambda) y)}{t}$$

(as f is convex). Q.E.D.

Jensen's Inequality: If $0 \leq \lambda_i \leq 1$, $\sum_{i=1}^{r} \lambda_i = 1$, and f is convex then

$$f\left(\sum_{i=1}^{r} \lambda_i x_i\right) \leq \sum_{i=1}^{r} \lambda_i f(x_i)\,.$$

Applying Jensen's Inequality with $f(x) = \binom{x}{t}$, $\lambda_i = 1/r$, $1 \leq i \leq r$, yields:

Lemma 12.2: *If $\sum_{i=1}^{r} x_i = N$ then $\sum_{i=1}^{r} \binom{x_i}{t} \geq r \binom{N/r}{t}$.*

Proof of Theorem 12.1 (lower bound): Fix $G \subseteq A_1 \times A_2$, $|G| = e$, so that there are exactly $B(m, n, a, b; e)$ pairs (B_1, B_2), $B_1 \in [A_1]^a$, $B_2 \in [A_2]^b$, $B_1 \times B_2 \subseteq G$. For $i \in A_1$ set

(12.3) $$X(i) = \{j \in A_2 : (i, j) \in G\}\,.$$

Then

(12.4) $$e = |G| = \sum_{i \in A_1} |X(i)|\,.$$

For $B_2 \in [A_2]^b$ set

(12.5) $$Y(B_2) = \{i \in A_1 : i \cdot B_2 \subseteq G\}\,.$$

Then

$$\begin{aligned}(12.6)\qquad &\{(i, B_2) : i \in A_1, B_2 \in [X(i)]^b\}\\ &= \{(i, B_2) : B_2 \in [A_2]^b, i \in Y(B_2)\}\end{aligned}$$

so

$$\begin{aligned}(12.7)\qquad \sum_{B_2 \in [A_2]^b} |Y(B_2)| &= \sum_{i \in A_1} \binom{|X(i)|}{b}\\ &\geq m\binom{e/m}{b}\end{aligned}$$

by Lemma 12.2. Then

$$\begin{aligned}(12.8)\qquad B(m, n, a, b; e) &= \sum_{B_2 \in [A_2]^b} \binom{|Y(B_2)|}{a}\\ &\geq \binom{n}{b}\binom{m\binom{e/m}{b}\Big/\binom{n}{b}}{a},\end{aligned}$$

again by Lemma 12.2. Q.E.D.

We apply (12.1) to Theorem 12.1 to find bounds on $k_{ab}(m, n)$. Note

$$(12.9)\qquad \binom{n}{b}\binom{m\binom{e/m}{b}\Big/\binom{n}{b}}{a} > 0 \ \text{ iff } \ m\binom{e/m}{b}\Big/\binom{n}{b} > a - 1\,.$$

Let us concentrate on the cases $a = b$ fixed, $m = n \to \infty$.

COROLLARY 12.1: $a!^{2/a^2}\, n^{2-2/a}\big(1 + o(1)\big) \leq k_a(n) \leq (a - 1)^{1/a}\, n^{2-1/a}\big(1 + o(1)\big)$.

We may improve Corollary 12.1 when $a = 2$. Let $n = p^2 + p + 1$, where p is a prime power. There exists a projective plane (see, e. g., M. HALL [1967]) with n points and n lines. Label the points $P_1, \ldots, P_n$ and the lines $L_1, \ldots, L_n$ arbitrarily. Let $A = (a_{ij})$ where $a_{ij} = 1$ if $P_i \in L_j$, 0 otherwise. Since two points cannot lie on two lines, A does not contain a 2×2 submatrix of ones. Hence

$$(12.10)\quad k_2(p^2 + p + 1) \geq 1 + \sum a_{ij} = 1 + (p^2 + p + 1)(p + 1)$$

(see Exercise 3). For all n there exists $m = p^2 + p + 1$, $n\big(1 - o(1)\big) < m \leq n$. Therefore

$$(12.11)\qquad k_2(n) \geq k_2(m) \geq n^{3/2}\big(1 - o(1)\big)$$

and (T. KŐVÁRI, V. T. SÓS, P. TURÁN [1954])

$$(12.12)\qquad \lim_{n\to\infty} k_2(n)\, n^{-3/2} = 1\,.$$

W. G. BROWN [1966] constructs a graph with n points, $\frac{n^{5/3}}{2}(1 + o(1))$ edges, containing no complete 3×3 bipartite graph. This implies, by the proof of Corollary 12.4 below, $k_3(n/2) \geq \frac{n^{5/3}}{4}(1 + o(1))$. Thus

$$(12.13) \qquad n^{5/3}\, 2^{-1/3}(1 + o(1)) \leq k_3(n) \leq n^{5/3}\, 2^{1/3}(1 + o(1)).$$

It is not known if $\lim k_3(n)\, n^{-5/3}$ exists. It seems probable that

$$(12.14) \qquad k_a(n) = O(n^{2-1/a}) \ (?)$$

but this has not been shown for $a > 3$.

We improve the lower bound to Corollary 12.1 by probabilistic means. For notational convenience, we call an $a \times a$ submatrix consisting entirely of ones an "aone". If $A = (a_{ij})$ is an $n \times n$ matrix define $t(A)$ as the number of aones in A. Define $A^* = (a_{ij}^*)$ by switching a "1" in each aone of A to "0". If the same "1" is chosen for many aones it is switched only once. The "1" is chosen in an arbitrary manner (e. g., the "upper left corner" of the aone). Then A^* contains no aones and, since at most $t(A)$ "1"s here been switched,

$$(12.15) \qquad \sum a_{ij}^* \geq \sum a_{ij} - t(A).$$

Now let $\mathbf{A} = \mathbf{A}_p = (\mathbf{a}_{ij})$ be a random $n \times n$ matrix with Prob $(\mathbf{a}_{ij} = 1) = p$. Then

$$(12.16) \qquad \mathsf{E}(\sum \mathbf{a}_{ij}) = \sum \mathsf{E}(\mathbf{a}_{ij}) = n^2 p$$

$$(12.17) \qquad \mathsf{E}(t(\mathbf{A})) = \binom{n}{a}^2 p^{a^2}$$

since there are $\binom{n}{a}^2$ $a \times a$ submatrices, each of which is an aone with probability p^{a^2}. Hence, applying expectations to (12.15),

$$(12.18) \qquad \mathsf{E}(\sum \mathbf{a}_{ij}^*) \geq n^2 p - \binom{n}{a}^2 p^{a^2}.$$

For n large we approximate $\binom{n}{a}$ by $n^a/a!$. We set $p = (a-1)!^{2/(a^2-1)-2/(a+1)}$ to maximize the right-hand side of (12.18). Then there exists a matrix A^* containing no aones,

$$(12.19) \qquad \begin{aligned} k_a(n) &\geq \sum a_{ij}^* \geq \mathsf{E}(\sum a_{ij}^*) \\ &\geq (1 - a^{-2})\,(a-1)!^{\frac{2}{a^2-1}}\, n^{2-\frac{2}{a+1}}. \end{aligned}$$

In deriving (12.19) we have used a common tool of the probabilistic method. In general, we wish to find a "structure" (e. g., matrix, graph, coloring) which contains no "forbidden subsystem" (aone, small cycle, monochromatic subgraph). To do this, we first take a random structure that contains only a small number of forbidden subsystems. Then we adjust the structure so that all of the forbidden subsystems are destroyed and no new ones are created. We have used this method in Theorems 4.2, 11.2, and shall use it again in Theorem 13.2.

COROLLARY 12.2: *Let* $|A_1| = |A_2| = n$, $A_1 \cap A_2 = \emptyset$. *If* $G \subseteq A_1 \times A_2$, $|G| \geq \frac{n^2}{2}(1 - o(1))$, *there exists* $B_i \subseteq A_i$, $B_1 \times B_2 \subseteq G$, $|B_i| \geq (1 - o(1)) \log_2 n$.

Corollary 12.2 gives an analogue to the Ramsey result (5.2). Let $g(n)$ be the maximal integer t such that if an $n \times n$ bipartite graph is 2-colored there exists a monochromatic $t \times t$ bipartite graph. In any 2-coloring some color must be used at least $n^2/2$ times. Thus $g(n) \geq (1 - o(1)) \log_2 n$. A random 2-coloring contains an expected number $\binom{n}{t}^2 2^{1-t^2}$ of monochromatic $t \times t$ bipartite graphs. For $t = (2 + o(1)) \log_2 n$ this number is less than unity so

$$(12.20) \qquad (1 - o(1)) \log_2 n \leq g(n) \leq (2 + o(1)) \log_2 n.$$

It seems difficult to improve either bound on $g(n)$ by a constant factor.

COROLLARY 12.3 (P. ERDŐS and J. W. MOON [1964]): *Let* $f(m, n, a, b)$ = *the number of monochromatic* $B_1 \times B_2$, $|B_1| = a$, $|B_2| = b$ *in any 2-coloring of* $A_1 \times A_2$, $|A_1| = m$, $|A_2| = n$. *Then, for* a, b *fixed*

$$\lim_{m,n \to \infty} \frac{f(m, n, a, b)}{\binom{m}{a}\binom{n}{b}} = 2^{1-ab}.$$

PROOF: Let $h : A_1 \times A_2 \to \{1, 2\}$ be any coloring. Set $e_i = |h^{-1}(i)|$, $i = 1, 2$. Then $e_1 + e_2 = mn$ and the number of monochromatic $B_1 \times B_2$ is at least

$$(12.21) \qquad \sum_{i=1}^{2} \binom{n}{b} \binom{m \binom{e_i/m}{b} \Big/ \binom{n}{b}}{a} \geq 2 \binom{n}{b} \binom{m \binom{n/2}{b} \Big/ \binom{n}{b}}{a}$$

(since the lower bound of (12.2) is convex in e)

$$= \binom{m}{a}\binom{n}{b} 2^{1-ab}(1 - o(1)).$$

Conversely, let **h** be a random coloring. Then the expected number of monochromatic $B_1 \times B_2$ is $\binom{m}{a}\binom{n}{b} 2^{1-ab}$ and so for some h there are $\leq \binom{m}{a}\binom{n}{b} 2^{1-ab}$ monochromatic $B_1 \times B_2$. Q.E.D.

The analogous result for complete graphs is not known. In particular, let $f(n, a) =$ the number of monochromatic $[S]^2$, $|S| = a$, in any 2-coloring of $[n]^2$. A. W. GOODMAN [1959] has shown

$$\begin{aligned} f(2n, 3) &= n(n-1)(n-2)/3 \\ f(4n+1, 3) &= 2n(n-1)(4n+1)/3 \\ f(4n+3, 3) &= 2n(n+1)(4n-1)/3 \end{aligned} \tag{12.22}$$

completely determining $f(n, 3)$. For $a > 3$ an expected value argument easily gives $f(n, a) < \binom{n}{a} 2^{1-\binom{a}{2}}$. It appears likely that

$$\lim_{n\to\infty} \frac{f(n, a)}{\binom{n}{a}} = 2^{1-\binom{a}{2}} \quad (?) \tag{12.23}$$

but this has not been proven. (See Exercise 2.)

COROLLARY 12.4: *Let n be even (n odd is similar). If $G \subseteq [n]^2$, $|G| \geq e$ then the number of $B_1 \times B_2 \subseteq G$, $B_1 \cap B_2 = \varnothing$, $|B_1| = a$, $|B_2| = b$,* is at least

$$\binom{n/2}{b} \binom{\frac{n}{2}\binom{2e/n}{b} \Big/ \binom{n/2}{b}}{a}. \tag{12.24}$$

PROOF: Fix G. By Exercise 6 we decompose $[n] = A_1 + A_2$ so that $G \cap (A_1 \times A_2) \geq e/2$. The number of $B_1 \times B_2 \subseteq G$, $B_1 \in [A_1]^a$, $B_2 \in [A_2]^b$ is at least $B(n/2, n/2, a, b, e/2)$. Q.E.D.

We generalize Theorem 12.1 to k-graphs.

THEOREM 12.2: *Let $n \geq t$ be integers, and $0 \leq \alpha \leq 1$. Set $g_0 = g_0(n, \alpha, t) = \alpha$ and*

$$g_{i+1} = g_{i+1}(n, \alpha, t) = \binom{ng_i}{t} \Big/ \binom{n}{t}. \tag{12.25}$$

Let $A_1, \ldots, A_k$ be disjoint with $|A_1| = \ldots = |A_k| = n$. Then if $G \subseteq A_1 \times \ldots \times A_k$, $|G| \geq \alpha n^k$, the number of $B_1 \times \ldots \times B_k \subseteq G$, $|B_i| = t$, is at least $g_k \binom{n}{t}^k$.

PROOF: We first note that, by repeated applications of Lemma 12.1, $g_i(n, \alpha, t)$ is a convex function of α. We prove Theorem 12.2 by induction on k. It is trivial for $k = 1$. Suppose the theorem holds for $k - 1$ and fix $G \subseteq A_1 \times \ldots \times A_k$, $|G| \geq \alpha n^k$. For $x \in A_k$ set

$$d(x) = |\{(x_1, \ldots, x_{k-1}) : (x_1, \ldots, x_{k-1}, x) \in G\}| / n^{k-1}. \tag{12.26}$$

$$\alpha n^k \leq |G| = \sum_{x \in A_k} d(x) n^{k-1} \tag{12.27}$$

$$\alpha n \leq \sum_{x \in A_k} d(x). \tag{12.28}$$

Set

$$U = \{(B_1, \ldots, B_{k-1}, x) : B_i \in [A_i]^t, x \in B_k, \quad B_1 \cdots B_{k-1} \cdot \{x\} \subseteq G\}. \tag{12.29}$$

Each $x \in A_k$ induces $G_x \subseteq A_1 \times \cdots \times A_{k-1}$, $|G_x| = d(x) n^{k-1}$. So by induction

$$|\{(B_1, \ldots, B_{k-1}) : (B_1, \ldots, B_{k-1}, x) \in U\}| \geq g_{k-1}(n, d(x), t) \binom{n}{t}^{k-1}. \tag{12.30}$$

So

$$|U| \geq \binom{n}{t}^{k-1} \sum_{x \in A_k} g_{k-1}(n, d(x), t) \geq \binom{n}{t}^{k-1} n g_{k-1}(n, \alpha, t) \tag{12.31}$$

by Lemma 12.2. Set

$$y(B_1, \ldots, B_{k-1}) = |\{x : (B_1, \ldots, B_{k-1}, x) \in U\}|. \tag{12.32}$$

Then

$$\sum y(B_1, \ldots, B_{k-1}) = |U| \geq \binom{n}{t}^{k-1} n g_{k-1}(n, \alpha, t). \tag{12.33}$$

The number of $B_1 \times \ldots \times B_{k-1} \times B_k \subseteq G$, $B_i \in [A_i]^t$, is

$$\sum \binom{y(B_1, \ldots, B_{k-1})}{t} \geq \binom{n}{t}^{k-1} \binom{n g_{k-1}(n, \alpha, t)}{t} \tag{12.34}$$

(again by Lemma 2)

$$= \binom{n}{t}^k g_k(n, \alpha, t)$$

by (12.25). Q.E.D.

To apply Theorem 12.2 we note that if $\alpha n \gg t$, $\binom{\alpha n}{t} \Big/ \binom{n}{t} \sim \alpha^t$.

COROLLARY 12.5 (P. ERDŐS [1965]): *Let h be a 2-coloring of the elements of* $A_1\times \ldots \times A_k$, $|A_i| = n$. *Then there exists a monochromatic set* $B_1\times \ldots \times B_k$, $B_i \subseteq A_i$, $|B_i| = t = (\log_2 n)^{1/(k-1)}(1-o(1))$.

PROOF: We prove the stronger statement that if $G \subseteq A_1\times \ldots \times A_k$, $|G| \geq n^k/2$, then such $B_1, \ldots, B_k$ exist. Set $\alpha_i = g_i(n, 1/2, t)$. So $\alpha_0 = 1/2$, $\alpha_{i+1} = \binom{n\alpha_i}{t}\Big/\binom{n}{t}$. For $0 \leq i \leq k-2$, $\alpha_{i+1} > \frac{1}{2}\alpha_i^t \geq \alpha_i^{t+1}$ (details are suppressed). Thus $\alpha_{k-1} > 2^{-(t+1)^{k-1}} = n^{-1+o(1)} > t/n$. Hence $\alpha_k > 0$, $g_k > 0$, so, by Theorem 12.2, $B_1, \ldots, B_k$ exist. Q.E.D.

Corollary 12.5 is close to best possible. For let **h** be a random 2-coloring of $A_1\times \ldots \times A_k$. The expected number of monochromatic $B_1\times \ldots \times B_k$ is $\binom{n}{t}^k 2\cdot 2^{-t^k} < n^k 2^{-t^{k-1}} < 1$ for $t = (k\log_2 n)^{1/(k-1)}$. Thus there exists a coloring h with no monochromatic $B_1\times \ldots \times B_k$, $|B_i| > (k\log_2 n)^{1/(k-1)}$.

The corresponding result for RAMSEY's Theorem is not nearly as good. Let $R_k(n)$ be the maximal t such that any 2-coloring of $[n]^k$ yields a monochromatic $[S]^k$, $|S| \geq t$. (Note this is not the R defined in § 5.) From Section 5, $C_2 \log n < R_2(n) < C_2' \log n$ where $C_2 = (2\log 2)^{-1}$, $C_2' = 2(\log 2)^{-1}$. For $k = 3$

$$C_3 \log\log n < R_3(n) < C_3' (\log n)^{1/2} \tag{12.35}$$

and for $k \geq 4$

$$C_k \underbrace{\log\log\ldots\log}_{(k-1)} n < R_k(n) < C_k' \underbrace{\log\log\ldots\log}_{(k-2)} n. \tag{12.36}$$

Moreover, for $k \geq 3$, $R_{k+1}(n) < C_k'' \log(R_k(n))$. Thus, if

$$R_3(n) < C\log\log n$$

could be shown then $R_k(n)$ would be known within a constant factor. The determination of the order of $R_3(n)$ is a major unsolved problem in this area.

We have not been able to improve the bounds on $R_k(n)$. However, we can use Theorem 12.2 to find a subset that is "substantially" one-color. Set $R_k(n, \varepsilon)$ = the maximal t such that any two-coloring of $[n]^k$ yields a set $S \subseteq [n]$, $|S| \geq t$, such that $\left(\frac{1}{2}+\varepsilon\right)$ of the k-edges of S are one-color. So $R_k(n, 1/2) = R_k(n)$. $R_k(n, \varepsilon)$ is monotone in n, ε. Bounds on $R_2(n, \varepsilon)$ are given by Corollary 7.2.

THEOREM 12.3: *For* $k \geq 3$ *there exist* $\varepsilon'(k)$, $c(k)$ *so that for* n *large*

$$R_k(n, \varepsilon'(k)) > c(k)(\log_2 n)^{1/(k-1)}.$$

PROOF: Fix k and let n be large. Let h be a two-coloring of $[n]^k$. For $D_1, \ldots, D_s$ disjoint subsets of $[n]$, and $a_1, \ldots, a_s$ nonnegative integers, $\sum a_i = k$, we define

$$h(D_1^{a_1} \ldots D_s^{a_s}) = \sum h(B) \tag{12.37}$$

over those $B \in [n]^k$, $|B \cap D_i| = a_i$. We let $h(S)$ denote $h(S^k)$. Let $A_1, \ldots, A_k$ be disjoint subsets, $|A_i| = [n/k]$, of $[n]$. Applying Corollary 12.5 we find, and fix, $B_1, \ldots, B_k$,

$$|B_i| = t = (\log_2(n/k))^{1/(k-1)}(1 - o(1)),$$

such that $|h(B_1 \cdots B_k)| = t^k$. We want to transform this into a large $h(S)$.

For $1 \leq i \leq k$ let $\mathbf{B}_i$ be a subset of B_i such that each $x \in B_i$ is in $\mathbf{B}_i$ with probability p_i. The p_i are constants, $0 \leq p_i \leq 1$, whose values will be determined later. Then

$$h(\mathbf{B}_1 \cup \ldots \cup \mathbf{B}_k) = \sum h(\mathbf{B}_1^{a_1} \ldots \mathbf{B}_k^{a_k}), \tag{12.38}$$

the summation over nonnegative a_i, $\sum a_i = k$. For any $a_1, \ldots, a_k$

$$\mathsf{E}[h(\mathbf{B}_1^{a_1} \ldots \mathbf{B}_k^{a_k})] = \sum_W h(W) \text{ Prob } [W \subseteq \mathbf{B}_1 \cup \ldots \cup \mathbf{B}_k], \tag{12.39}$$

the summation over $W \in [n]^k$, $|W \cap B_i| = a_i$.

$$\text{Prob } [W \subseteq \mathbf{B}_1 \cup \ldots \cup \mathbf{B}_k] = \text{Prob } [W \cap B_i \subseteq \mathbf{B}_i \text{ for } 1 \leq i \leq k] = \Pi p_i^{a_i} \tag{12.40}$$

independent of W. Applying (12.40) to (12.39),

$$\mathsf{E}[h(\mathbf{B}_1^{a_1} \ldots \mathbf{B}_k^{a_k})] = (\Pi p_i^{a_i}) \sum_W h(W) = (\Pi p_i^{a_i}) h(B_1^{a_1} \ldots B_k^{a_k}). \tag{12.41}$$

Applying expectations to (12.38),

$$\mathsf{E}[h(\mathbf{B}_1 \cup \ldots \cup \mathbf{B}_k)] = \sum h(B_1^{a_1} \ldots B_k^{a_k})(\Pi p_i^{a_i}). \tag{12.42}$$

We want to adjust the values of p_i so that $\mathsf{E}[h(\mathbf{B}_1 \cup \ldots \cup \mathbf{B}_k)]$ is large. To do this we require a result in polynomial approximations.

Let T_k be the set of real polynomials $G(x_1, \ldots, x_k) = \sum C_{a_1 \ldots a_k} \Pi x_i^{a_i}$, where the summation is over nonnegative integral $a_i \geq 0$, $\sum a_i = k$. Set $|G| = \max |C_{a_1 \ldots a_k}|$, $\|G\| = \max_{0 \leq x_i \leq 1} |G(x_1, \ldots, x_k)|$.

LEMMA 12.3: *There exists* $\varepsilon = \varepsilon(k)$ *so that* $\|G\| \geq \varepsilon |G|$ *for all* $G \in T_k$.

PROOF: T_k is a metric space under norm $|\cdot|$. Set

$$T'_k = \{G : |G| = 1\}.$$

Then T'_k is compact. If $G \in T'_k$, $G \neq 0$, so $||G|| > 0$. The function $||\cdot||$ is continuous and positive on the compact set T'_k so there exists $\varepsilon = \varepsilon(k)$, $||G|| \geq \varepsilon$ for $G \in T'_k$.

If $G = 0$, $||G|| \geq \varepsilon |G|$ trivially. If $G \neq 0$, $G = |G| G_1$, $G_1 \in T'_k$. Then, since $||\cdot||$ is linear,

$$||G|| = |G| \, ||G_1|| \geq \varepsilon |G|.$$

Q.E.D.

It is possible, though we have not done so, to give an explicit form for $\varepsilon(k)$.

By Lemma 12.3 we find, and fix, $p_1, \ldots, p_k$ so that, by (12.42),

$$(12.43) \qquad |\mathsf{E}[h(\mathbf{B}_1 \cup \ldots \cup \mathbf{B}_k)]| \geq |h(B_1 \cdots B_k)| \, \varepsilon(k) = \varepsilon(k) t^k.$$

Thus, there exist fixed sets $B'_1, \ldots, B'_k$, $S = B'_1 \cup \ldots \cup B'_k$

$$(12.44) \qquad |h(S)| \geq \varepsilon(k) t^k.$$

We know $|S| \leq |B_1 \cup \ldots \cup B_k| = kt$. As $|h(S)| \leq \binom{|S|}{k}$, $|S| \geq$ $\geq [k!\,\varepsilon(k)]^{1/k} t$. Setting $c(k) < [k!\,\varepsilon(k)]^{1/k}$, $\varepsilon'(k) > \varepsilon(k) k!\, k^{-k}/2$, Theorem 12.3 is satisfied by S. Q.E.D.

Theorem 12.3 leads to an intriguing problem. Take $k = 4$, as an example. For ε small $R_4(n, \varepsilon)$ is of the order $(\log n)^{1/3}$. (The reader may check by taking a random coloring that for ε fixed, $\varepsilon > 0$, $R_k(n, \varepsilon) < c(k, \varepsilon) (\log n)^{1/(k-1)}$). For $\varepsilon = 1/2$, $R_4(n, \varepsilon) = R_4(n) = 0(\log \log n)$. How does the function $R_4(n, \varepsilon)$ change as ε varies? Is there a "cut off point" ε_0 so that for $\varepsilon < \varepsilon_0$, $R_4(n, \varepsilon)$ is of order $(\log n)^{1/3}$ but for $\varepsilon > \varepsilon_0$, $R_4(n, \varepsilon)$ is of order $\log \log n$? What is the order of $R_4(n, .5 - \varepsilon)$ for ε small? We cannot answer these questions.

Exercises

1. Using the notation of Theorem 12.1 show

$$B(m, n, a, b; e) \leq \binom{m}{a} \binom{n}{b} \frac{\binom{mn - ab}{e - ab}}{\binom{mn}{e}}.$$

Compare the bounds on $B(m, n, a, b; e)$ for m, n, e "much larger" than a, b.

2. Show that $f(n, a) > \binom{n}{4^a} \Big/ \binom{n-a}{4^a-a}$ so that for a fixed, $f(n, a) > \binom{n}{a}(a!\,4^{a^2})\big(1 - o(1)\big)$. (Hint: Ramsey Number $R(a, a) < 4^a$.) Show that the limit in (12.23) exists.

Much work (see R. GUY and S. ZNÁM [1969] for summary) has been done on finding the exact value $k_{ab}(m, n)$ for small values of a, b, m, n. Lemma 12.2 is replaced by (*) "If $\sum x_i = N$, $\sum \binom{x_i}{t}$ is minimized by taking x_i as nearly equal as possible." and we apply this modification to (12.9).

3. (I. REIMAN [1958]). Prove

$$k_2(p^2 + p + 1) \leq 1 + (p^2 + p + 1)(p + 1).$$

This implies equality in (12.10).

4. Prove $k_2(5) \leq 13$.

5. Prove $k_2(9) \leq 30$. (This requires some combinatorial manipulations.)

6. (a) Let $G \subseteq [n]$. Prove there exists a decomposition $[n] = A_1 + A_2$ so that

$$|G \cap (A_1 \times A_2)| \geq |G|/2.$$

(Hint: Let $[n] = \mathbf{A}_1 + \mathbf{A}_2$. Note we get a slight improvement if $\mathbf{A}_1$ is a random member of $[n]^{[n/2]}$.)

(b) Let $G \subseteq [n]^r$. Prove there exists a decomposition $[n] = A_1 + \ldots + A_r$ so that

$$|G \cap (A_1 \times \ldots \times A_r)| \geq |G|\, r!\, r^{-r}.$$

7. Fix r, integral, and c, $0 < c < 1$. Let $f(n) =$ the maximal integer such that if $G \subseteq [n]^2$, $|G| \geq c\binom{n}{2}$, then there exists an $r \times f(n)$ bipartite graph in G. Prove

$$(*) \qquad f(n) = \big(1 + o(1)\big)n2^{-r}.$$

Widen the conditions on r, c so that (*) still holds. (E. g.: $r < K \log n$, $c > n^{-\varepsilon}$.)

8. Fix α, $0 < \alpha < 1$. Let $f(n)$ be the maximal integer such that if A is an $n \times n$ $0 - 1$ matrix with at least αn^2 ones there exists an $r \times s$ all 1 submatrix with $rs \geq f(n)$. Find $f(n)$ within a $\big(1 + o(1)\big)$ factor. [This problem is due to G. A. REID.]

13. PACKING, COVERING AND TURÁN'S THEOREM

Let $2 \leq l \leq k \leq n$ be positive integers. We will denote by $T(n, k, l)$ — T for TURÁN — the smallest q such that there exists an l-graph with n vertices, q edges, and with no independent set of size k. In other words, $T(n, k, l)$ is the smallest number of l-subsets of an n-set V such that every k-subset of V contains at least one of these subsets.

For $l = 2$ the value of T is given by the following famous theorem of TURÁN.

TURÁN'S THEOREM [1941]: *Let $n = (k-1)m + a$, $0 \leq a < k-1$. Then*

$$T(n, k, 2) = (k-1)\binom{m}{2} + ma \,. \tag{13.1}$$

We suppress the proof, as it does not involve probabilistic considerations.

TURÁN's theorem is often stated as follows: Let $f(n, k)$ denote the maximal number of edges in a graph on n points containing no clique of size k. Then

$$f(n, k) = \binom{n}{2} - (k-1)\binom{m}{2} - ma \tag{13.2}$$

where m, a are defined as before. The two statements of TURÁN's theorem are equivalent since a graph contains no clique of size k iff its complement contains no k independent points.

TURÁN's Theorem is the first instance of a problem in "extremal graph theory". The general problem is to find, for a given graph H, the maximal $|G|$ of a graph G on n points not containing H. The case $H = K_{ab}$ has been examined in §12.

The generalization of the function T to larger l was first suggested by P. TURÁN [1941, 1961]. V. CHVÁTAL [1970] contains a survey of known results on $T(n, k, l)$. A construction due to TURÁN gives the upper bound

$$T(n, k, l) \leq \binom{n}{l}\left[\frac{k-1}{l-1}\right]^{1-l} \tag{13.3}$$

(see V. CHVÁTAL [1970] or G. KATONA, T. NEMETZ and M. SIMONOVITS [1964]). A lower bound to T is given by the following result.

THEOREM 13.1 (G. KATONA, T. NEMETZ and M. SIMONOVITS [1964]):

$$T(n, k, l) \geq \binom{n}{l} \bigg/ \binom{k}{l}. \tag{13.4}$$

PROOF: Let G be a minimal l-graph on $[n]$ with no k independent points. Each l-edge of G is contained in exactly $\binom{n-l}{k-l}$ k-sets. Since each k-set contains some l-edge

$$|G| \binom{n-l}{k-l} \geq \binom{n}{k}. \tag{13.5}$$

Therefore,

$$T(n, k, l) = |G| \geq \binom{n}{k} \bigg/ \binom{n-l}{k-l} = \binom{n}{l} \bigg/ \binom{k}{l}. \tag{13.6}$$

Q.E.D.

We restate the proof of Theorem 13.1 in probabilistic terms. Let G be fixed and $\mathbf{A}$ be a random member of $[n]^k$. Then the expected number of l-edges of G contained in $\mathbf{A}$ is $|G| \binom{n-l}{k-l} \Big/ \binom{n}{k}$. If this number is < 1 then some A is independent, a contradiction. This yields Theorem 13.1 and points the way to the following result.

THEOREM 13.2 (J. SPENCER [1972a]): *If* $n \geq (k-1)l/(l-1)$,

$$T(n, k, l) \geq \left(\frac{n}{l}\right)^l \left(\frac{k-1}{l-1}\right)^{1-l}.$$

PROOF: Fix $G \subseteq [n]^l$, $|G| = T(n, k, l)$, with no k independent points. Let $\mathbf{A} = \mathbf{A}_p =$ a random subset of $[n]$ where for each $i \in [n]$

$$\text{Prob}\,[i \in \mathbf{A}] = p$$

and these probabilities are independent. Here p is a constant whose value will be determined later.

$$\mathsf{E}(|\mathbf{A}|) = \sum_{i \in [n]} \text{Prob}\,[i \in \mathbf{A}] = pn \tag{13.7}$$

$$\mathsf{E}(|G \cap [\mathbf{A}]^k|) = \sum_{B \in G} \text{Prob}\,[B \subseteq \mathbf{A}]$$

$$= |G|\, p^l.$$

Now fix a "choice" function v that assigns to each nonempty subset $S \subseteq [n]$ an element $v(S) \in S$. Then if $A \subseteq [n]$

$$A^* = A - \bigcup_{B \in [A]^k \cap G} \{v(B)\} \tag{13.8}$$

is an independent set. Since

$$|A^*| \geq |A| - |G \cap [A]^k| \tag{13.9}$$

$$\mathsf{E}[|\mathbf{A}^*|] \geq pn - |G|p^l. \tag{13.10}$$

But every $|A^*| \leq k-1$, since A^* is independent. Consequently

$$k - 1 \geq \mathsf{E}[|\mathbf{A}^*|] \geq pn - |G|p^l \tag{13.11}$$

$$|G| \geq (pn - (k-1))p^{-l}. \tag{13.12}$$

Since $n \geq (k-1)l/(l-1)$, we set $p = (k-1)\left(\frac{l}{l-1}\right)\Big/n \leq 1$, yielding Theorem 13.2. Q.E.D.

We combine the methods of Theorems 13.1, 13.2 to get the following result.

THEOREM 13.3: *For* $k \leq a \leq n$,

$$T(n, k, l) \geq (a - (k-1))\binom{n}{l}\Big/\binom{a}{l}$$

PROOF: Fix $G \subseteq [n]^l$, $|G| = T(n, k, l)$, with no k independent points. Let $\mathbf{A} = \mathbf{A}^a$ be a random member of $[n]^a$.

$$\mathsf{E}(|\mathbf{A}|) = a \tag{13.13}$$

$$\mathsf{E}(|G \cap [\mathbf{A}]^k|) = \sum_{B \in G} \text{Prob}\,[B \subseteq \mathbf{A}] = |G|\binom{a}{l}\Big/\binom{n}{l}. \tag{13.14}$$

Define A^* by (13.8). Then

$$k - 1 \geq \mathsf{E}(|\mathbf{A}^*|) \geq a - |G|\binom{a}{l}\Big/\binom{n}{l} \tag{13.15}$$

which yields Theorem 13.3. Q.E.D.

Another proof of Theorem 13.3 is given in Exercise 4.

Setting $a = k$ in Theorem 13.3 yields Theorem 13.1. We claim that Theorem 13.3 always gives an improvement over (or at least the same result as) Theorem 13.2. That is

$$\max_{k \leq a \leq n} (a - (k-1))\binom{n}{l}\Big/\binom{a}{l} \geq n^l(k-1)^{1-l}(l-1)^{l-1}l^{-l}. \tag{13.16}$$

We first consider a more general result. Let f be any real valued function with domain the family of all subsets of $[n]$. Then

$$\begin{aligned}\mathsf{E}[f(\mathbf{A}_p)] &= \sum_{a=0}^{n} \text{Prob}\,(|\mathbf{A}_p| = a)\, \mathsf{E}[f(\mathbf{A}_p) \mid |\mathbf{A}_p| = a] \\ &= \sum_{a=0}^{n} \text{Prob}\,(|\mathbf{A}_p| = a)\, \mathsf{E}[f(\mathbf{A}^a)].\end{aligned} \tag{13.17}$$

Here we used that the conditional distribution of $\mathbf{A}_p$ given $|\mathbf{A}_p| = a$ is $\mathbf{A}^a$. Since $\sum_{a=0}^{n} \text{Prob}(|\mathbf{A}_p| = a) = 1$,

$$\text{(13.18)} \qquad \min \mathsf{E}[f(\mathbf{A}^a)] \leq \mathsf{E}[f(\mathbf{A}_p)] \leq \max \mathsf{E}[f(\mathbf{A}^a)].$$

We can now prove (13.16). Fix k, l, $n \geq (k-1)l/(l-1)$, $p = (k-1)\left(\frac{l}{l-1}\right)\Big/n$. Set

$$\text{(13.19)} \qquad f(A) = |A| - \big(pn - (k-1)\big)p^{-l}\binom{|A|}{l}\Big/\binom{n}{l}.$$

We require

$$\text{(13.20)} \qquad \mathsf{E}\left[\binom{|\mathbf{A}_p|}{l}\right] = \sum_{B \in [n]^l} \text{Prob}\,[B \subseteq \mathbf{A}_p] = \binom{n}{l}p^l.$$

By (13.18) there exists a, $0 \leq a \leq n$,

$$\text{(13.21)} \qquad \begin{aligned} a - \big(pn - (k-1)\big)p^{-l}\binom{a}{l}\Big/\binom{n}{l} &= \mathsf{E}[f(\mathbf{A}^a)] \\ > \mathsf{E}[f(\mathbf{A}_p)] = pn - \big(pn - (k-1)\big)p^{-l}p^l &= k - 1 \end{aligned}$$

which implies (13.16).

In many problems one has a choice between using $\mathbf{A}_p$ and $\mathbf{A}^i$. In graph-theory problems the choice is between $\mathbf{G}_{n,p}$ and a random graph on n points with i edges. We prefer to use $\mathbf{A}_p$ and $\mathbf{G}_{n,p}$ as the answers tend to be expressed in exponentials instead of binomial coefficients and are therefore more tractible. However, as we have indicated, use of $\mathbf{A}^i$ will give a better result. Therefore, it should be used where exact answers are desired—especially finite problems.

So far, we have thought of $T(n, k, l)$ in a graph-theoretic sense. Thought of combinatorially, T becomes one of four functions to investigate.

Definitions: Let $n \geq k \geq l \geq 2$.

$$T(n, k, l) = \min |F| : F \subseteq [n]^l,\ S \in [n]^k \text{ implies } \exists A \in F,\ A \subseteq S$$

$$t(n, k, l) = \max |F| : F \subseteq [n]^l,\ (A_1, A_2 \in F,\ A_1, A_2 \subseteq S \in [n]^k) \text{ implies } A_1 = A_2$$

$$M(n, k, l) = \min |F| : F \subseteq [n]^k,\ S \in [n]^l \text{ implies } \exists A \in F,\ A \supseteq S$$

$$m(n, k, l) = \max |F| : F \subseteq [n]^k,\ (A_1, A_2 \in F,\ A_1, A_2 \supseteq S \in [n]^l) \text{ implies } A_1 = A_2.$$

A more intuitive sense of these definitions may be gained by saying that a set A covers a set B if $A \subseteq B$ or $B \subseteq A$. Then fix $[n]$ as the universal set. $T(n, k, l)$ is the minimal number of l-sets to cover all k-sets; $t(n, k, l)$ is the maximal number of l-sets to cover no k-set more than once; $M(n, k, l)$ is the minimal number of k-sets to cover all l-sets; $m(n, k, l)$ is the maximal number of k-sets to cover no l-set more than once.

Theorem 13.1 may easily be generalized to show

$$(13.22)\qquad \begin{aligned} t(n, k, l) &\leq \binom{n}{l}\Big/\binom{k}{l} \leq T(n, k, l) \\ m(n, k, l) &\leq \binom{n}{l}\Big/\binom{k}{l} \leq M(n, k, l)\ . \end{aligned}$$

Note that $t(n, k, l) = 1$ if $2l \leq k$. The functions M, m have been studied by J. SCHÖNHEIM [1966].

Definition: An (n, k, l) tactical configuration (t. c.) is a family $F \in [n]^k$ so that every l-set is covered by exactly one member of F.

If such a t. c. exists, $M(n, k, l) = m(n, k, l) = \binom{n}{l}\Big/\binom{k}{l}$. There has been much study of the existence and construction of tactical configurations. [E. g., the $(n, 3, 2)$ t. c. are called STEINER Systems, first studied in 1847 (see M. HALL [1967])]. Counting arguments (H. HANANI [1960]) show that if an (n, k, l) tactical configuration exists then

$$(13.23)\qquad \binom{n-l+t}{t}\Big/\binom{k-l+t}{t} = \text{integer, for } t = 1, \ldots, l.$$

Fixing k, l (13.23) implies that n satisfies a certain congruence condition. H. HANANI [1960, 1961] has shown that (13.23) is sufficient for the (n, k, l) t. c. to exist for $(k, l) = (3, 2), (4, 2), (5, 2), (4, 3)$. However, (13.23) is not always sufficient. For example, an $(n^2 + n + 1, n + 1, 1)$ tactical configuration is equivalent to a projective plane of order n. The conditions (13.23) are satisfied but the famous theorem of R. H. BRUCK and H. J. RYSER [1949] gives an infinite class of n for which the projective planes do not exist.

Evidence suggests that for k, l fixed and n approaching infinity, while satisfying (13.22) the number of nonisomorphic (n, k, l) t. c., approaches infinity very rapidly. T. DOYEN [1970] has proven this for the Steiner Triple case, $k = 3$, $l = 2$. It is conceivable that such designs could be constructed by means of probabilistic methods. A new idea would probably be required because the designs to be constructed are "precise" objects. We note that the construction of other kinds of designs, e. g., Hadamard matrices, also might be possible using probabilistic ideas. Unfortunately, the authors' brains are too old and too, weak respectively, to discover this construction.

P. Erdős and H. Hanani [1963] conjecture that for k, l fixed

$$(13.24)\quad \lim_n M(n, k, l) \binom{k}{l} \Big/ \binom{n}{l} = \lim_n m(n, k, l) \binom{k}{l} \Big/ \binom{n}{l} = 1.$$

(See Exercises 1 and 2.) By constructive methods, they prove (13.24) for $l = 2$, all k, and for $l = 3$, $k = p$ or $p + 1$ where p is a prime power. This conjecture also seems a likely candidate for a probabilistic proof, though we are presently unable to give one. Probabilistic methods give the following upper bound on M.

Theorem 13.4: $M(n, k, l) \leq \left[\binom{n}{l} \Big/ \binom{k}{l}\right] \left[1 + \log \binom{k}{l}\right]$.

Proof: We need to find an $F \subseteq [n]^k$ that covers all l-sets. Let $\mathbf{S}_1$, $\mathbf{S}_2, \ldots$ be random members of $[n]^k$ and set $\mathbf{F}_x = \{\mathbf{S}_1, \ldots, \mathbf{S}_x\}$. Set $u(\mathbf{F}_x) = \{A \in [n]^l : A \not\subseteq \mathbf{S}_i,\ 1 \leq i \leq x\}$, the uncovered l-sets. Since

$$\text{Prob}\,[A \in u(\mathbf{F}_x)] = \left[1 - \binom{k}{l} \Big/ \binom{n}{l}\right]^x,$$

(13.25)

$$\mathsf{E}[|u(\mathbf{F}_x)|] = \binom{n}{l} \left[1 - \binom{k}{l} \Big/ \binom{n}{l}\right]^x.$$

Let $v : [n]^l \to [n]^k$ be a "choice" function so that $v(A) \subseteq A$. Then, for any F,

$$(13.26)\quad F^* = F \cup \{v(A) : A \in u(F)\}$$

covers all l-sets.

$$(13.27)\quad \mathsf{E}[|\mathbf{F}_x^*|] \leq x + \binom{n}{l} \left[1 - \binom{k}{l} \Big/ \binom{n}{l}\right]^x.$$

Set $x = \left[\binom{n}{l} \Big/ \binom{k}{l}\right] \log \binom{k}{l}$ (see Exercise 3.) Pick F_x so that $|F_x^*| \leq$ $\leq \mathsf{E}[|\mathbf{F}_x^*|] \leq \left[\binom{n}{l} \Big/ \binom{k}{l}\right] \left[1 + \log \binom{k}{l}\right]$. Q.E.D.

A correspondingly good lower bound for $m(n, k, l)$ appears difficult. The best we can show is

$$(13.28)\quad m(n, k, l) \geq T(n, k, l) \Big/ \binom{k}{l}.$$

To show (13.28) note that a family of $m(n, k, l)$ k-sets covers $m(n, k, l) \binom{k}{l}$ l-sets. Since no more k-sets may be added without covering some l-set twice, the l-sets must cover all the k-sets. Thus $m(n, k, l) \binom{k}{l} \geq T(n, k, l)$.

Exercises

1. Prove $\lim_n M(n, k, l) \binom{k}{l} \Big/ \binom{n}{l}$, $\lim_n m(n, k, l) \binom{k}{l} \Big/ \binom{n}{l}$ exist. (Hint: For M it suffices to show $M(n, k, l) \binom{k}{l} \Big/ \binom{n}{l}$ is monotonically decreasing in n. Let $n_1 < n_2$, $F \subseteq [n_2]^l$ cover all l-sets. Then for every $S \in [n_2]^{n_1}$, $F \cap [S]^k \geq M(n_1, k, l)$.)

2. Prove $\lim_n M(n, k, l) \binom{k}{l} \Big/ \binom{n}{l} = 1$ iff $\lim_n m(n, k, l) \binom{k}{l} \Big/ \binom{n}{l} = 1$.

3. The proof of Theorem 13.3 is not exact because $x = \left[\binom{n}{l} \Big/ \binom{k}{l}\right] \log \binom{k}{l}$ is not an integer. To obtain the precise result replace $\mathbf{F}_x$ by $\mathbf{F}_p \subseteq [n]^k$ so that $\text{Prob}\,[A \in \mathbf{F}_p] = p = x \Big/ \binom{n}{k}$.

4. (G. Katona, T. Nemetz and M. Simonovits [1964]). Prove
(a) $T(n, k, l) \geq n - (k - 1)$.
(b) $T(n, k, l) \binom{n}{l}^{-1}$ is monotone increasing in n.
Use (1), (2) to prove Theorem 13.3.

5. Let $n \geq l$ be integers, $0 \leq p \leq 1$. Prove

$$\sum_{a=0}^{n} \left[\binom{n}{a} p^a (1 - p)^{n-a}\right] \binom{a}{l} = \binom{n}{l} p^l .$$

$\left[\text{Hint: Apply (13.17) with } f(A) = \binom{|A|}{l}.\right]$

6. Prove

$$n\, 2^n \leq T(2n^2, n^2, n) \leq n^2\, 2^n \big(2 \log 2 + o(1)\big) .$$

7. Prove

$$T(2r, r, k) \geq m(k)$$

where m is defined as in § 4.

8. (P. Erdős and A. Hajnal [1968]). Let $f : 2^{[n]} \to [n]$ such that for all $X \subseteq [n]$, $f(X) \notin X$. Say S is independent if $X \subseteq S$ implies $f(X) \notin S$. Set $g(n) =$ the maximal integer such that given any such set mapping f there exists an independent set S, $|S| \geq g(n)$. Prove

$$\log_2 n - c_1 \log \log n \leq g(n) \leq \log_2 n + c_2 \log \log n .$$

9. Let $f : 2^{[n]} \to [n]$. Call a set $X \subseteq [n]$ complete if

$$\bigcup_{S \subseteq X} f(S) = [n]$$

Call f r-complete if $|X| \geq r$ implies X is complete. Let $g(n)$ be the minimal integer such that there exists an r-complete $f : 2^{[n]} \to [n]$. Prove

$$\log_2 n \leq g(n) \leq \log_2 n + \log_2 \log_2 n + O(1)$$

It is not known if

$$g(n) = \log_2 n + O(1) \ (?) .$$

14. ASYMMETRIC GRAPHS

An automorphism of a graph G is a permutation σ of the vertices of G such that $G = \sigma(G)$. That is, $\{x, y\} \in G$ iff $\{\sigma(x), \sigma(y)\} \in G$. A graph is said to be symmetric if it contains an automorphism $\sigma \neq$ $\neq$ identity. A graph is asymmetric if it is not symmetric.

We review some elementary results in permutations on n points. Every permutation σ has a cycle representation

$$\sigma = (b_{11}, \ldots, b_{1c_1})\,(b_{21}, \ldots, b_{2c_2}) \ldots (b_{r1}, \ldots, b_{rc_r}) \tag{14.1}$$

which implies that

$$\sigma(b_{ij}) = b_{i,j+1}\ (1 \leq j \leq c_i, \bmod c_i)\,.$$

If σ has a_k cycles of length k we say that σ has cycle type

$$1^{a_1}\, 2^{a_2} \ldots n^{a_n}\,. \tag{14.2}$$

The number of such σ is

$$\frac{n!}{\prod_{k=1}^{n} a_k!\,k^{a_k}}\,. \tag{14.3}$$

If a graph G is asymmetric it can be made symmetric by adding and/or deleting certain edges. The asymmetry of G, denoted by $A(G)$, is defined as the minimal number of edges that need be added and/or deleted to make G symmetric. That is, $A(G) = \min |D|$ where D is a set of edges such that $G \,\Delta\, D$ is symmetric.

THEOREM 14.1 (P. ERDŐS and A. RÉNYI [1963]): *Let $G = G_n$ be a graph on n vertices. Then*

$$A(G) \leq (n-1)/2\,.$$

PROOF: For $1 \leq x \leq n$ set $S_x = \{y : \{x, y\} \in G\}$.

$$\begin{aligned}
\sum_{\{x,y\}} |S_x \,\Delta\, S_y - \{x, y\}| &= \sum_z |\{\{x, y\} : z \neq x, y;\, z \in S_x,\, z \notin S_y\}| \\
&= \sum_z |S_z|\,(n - 1 - |S_z|) \qquad (14.4)\\
&\leq n(n-1)^2/4\,.
\end{aligned}$$

Since the left-hand side of (14.4) has $n(n-1)/2$ addends, some $|S_x \Delta S_y - \{x, y\}| \leq (n-1)/2$. Fix x, y. Set

$$D = \{\{x, z\} : z \neq x, y; z \in S_x \Delta S_y\}. \tag{14.5}$$

Then $G \Delta D$ has the automorphism $\sigma = (xy)$ (where we delete cycles of length one for simplicity).

$$|D| \leq |S_x \Delta S_y - \{x, y\}| \leq (n-1)/2. \tag{14.6}$$

THEOREM 14.2 (P. ERDŐS and A. RÉNYI [1963]): *If $C > 1$ and n is sufficiently large there exists a graph G on n vertices such that*

$$A(G) \geq \frac{n}{2} - C\sqrt{n \log n}.$$

PROOF: Let $\mathbf{G} = \mathbf{G}_{n,1/2}$ be a random graph on n points. We shall show

$$\text{Prob}\left[\exists \sigma \neq 1 \exists D, |D| \leq \frac{n}{2} - C\sqrt{n \log n}, \sigma(\mathbf{G}\Delta D) = \mathbf{G}\Delta D\right] < 1. \tag{14.7}$$

For each $\sigma \neq 1$ set $m(G, \sigma) =$ the minimal $|D|$ where D is a set of edges such that $\sigma(G \Delta D) = G \Delta D$. Set

$$P(\sigma) = \text{Prob}\left[m(\mathbf{G}, \sigma) \leq \frac{n}{2} - C\sqrt{n \log n}\right]. \tag{14.8}$$

It suffices to show

$$\sum_{\sigma \neq 1} P(\sigma) < 1. \tag{14.9}$$

To prove (14.9) we shall bound $P(\sigma)$ for $\sigma \neq 1$. First, we require a condition for $\sigma(G \Delta D) = G \Delta D$. Suppose $\sigma \neq 1$ has cycle type $1^{a_1} 2^{a_2} \ldots n^{a_n}$. We define graph $H(\sigma)$. Its vertices are the $\binom{n}{2}$ elements $e \in [n]^2$. Its edges are defined by $\{e_1, e_2\} \in H(\sigma)$ iff $e_2 = \sigma(e_1)$.

$e = \{i, j\}$ is an isolated point of $H(\sigma)$ iff $e = \sigma(e)$. This occurs iff $\sigma(i) = i$ and $\sigma(j) = j$ $\left[\text{which is true for exactly } \binom{a_1}{2} \ e\text{'s}\right]$ or if $\{i, j\}$ is a 2-cycle (which is true for exactly a_2 e's). Te remaining $\binom{n}{2} - \binom{a_1}{2} - a_2$ points all have degree 2 and therefore form disjoint cycles.

Relabel the nonfixed members of $[n]^2$ as $e_{ij}(1 \leq i \leq t, 1 \leq j \leq d_i)$ such that $(e_{ij}, e_{i,j+1}) \in H(\sigma)$ where $j+1$ is recalculated modulo d_i.

Then $\sigma(G \Delta D) = G \Delta D$ iff for $1 \leq i \leq t$ either all $e_{ij} \in G \Delta D$ or all $e_{ij} \notin G \Delta D$. The minimal D such that $\sigma(G \Delta D) = G \Delta D$ is given by setting, for $1 \leq i \leq t$,

(14.10) $$D \cap \{e_{ij}\} = \{e_{ij}\} \cap G \quad \text{or} \quad \{e_{ij}\} \cap G^c$$

whichever is minimal. Set

(14.11) $$\mathbf{X}_{ij} = \begin{cases} 1 & \text{if } \ e_{ij} \in \mathbf{G} \\ 0 & \text{otherwise.} \end{cases}$$

Then the $\mathbf{X}_{ij}$ are independent random variables and

(14.12) $$m(\mathbf{G}, \sigma) = \sum_{i=1}^{t} \min\left[\sum_{j=1}^{d_i} \mathbf{X}_{ij},\ d_i - \sum_{j=1}^{d_i} \mathbf{X}_{ij}\right].$$

Set

(14.13) $$\mathbf{Y}_i = \min\left[\sum_{i=1}^{d_i} \mathbf{X}_{ij},\ d_i - \sum_{j=1}^{d_i} \mathbf{X}_{ij}\right].$$

Now we place bounds on $P(\sigma)$. Let $q(\sigma) = n - a_1 = |\{i : \sigma(i) \neq i\}|$.

Case 1: $q(\sigma) = 2$. Then $\sigma = (ij)$ and $H(\sigma)$ is the union of $(n-2)$ 2-cycles $(\{i, x\}, \{j, x\}$ for $x \neq i, j)$. The $\mathbf{Y}_i$ are independent $0 - 1$ variables.

(14.14) $$P(\sigma) = \text{Prob}\left[\sum_{i=1}^{n-2} \mathbf{Y}_i < \frac{n}{2} - C\sqrt{n \log n}\right] = o(n^{-2})$$

since $C > 1$.

Case 2: $q(\sigma) = 3$. Then $\sigma = (ijk)$ and $H(\sigma)$ is the union of $(n-2)$ 3-cycles $(\{i, j\}, \{j, k\}, \{i, k\}$ and $\{i, x\}, \{j, x\}, \{k, x\}$ for $x \neq i, j, k)$. Then $\mathbf{Y}_i = 0$ or 1, Prob $(\mathbf{Y}_i = 1) = 3/4$ and the $\mathbf{Y}_i$ are independent.

(14.15) $$P(\sigma) = \text{Prob}\left[\sum_{i=1}^{n-2} \mathbf{Y}_i < \frac{n}{2} - C\sqrt{n \log n}\right] < (\sqrt{3}/2)^n.$$

Case 3: $q(\sigma) \geq 4$. As with the case $q(\sigma) = 3$, $P(\sigma)$ is "exponentially small" for $q(\sigma) \geq 4$. In fact (Exercise 6)

(14.16) $$P(\sigma) < n^{-q(\sigma)-2} \text{ for } q(\sigma) \geq 4.$$

The number of σ with $q(\sigma) = q$ is

(14.17) $$\binom{n}{q} q! \sum_{i=0}^{q} (-1)^i/i! \sim \frac{n!}{(n-q)!\,e} < n^q.$$

Thus

$$(14.18) \quad \sum_{\sigma \neq 1} P(\sigma) < n^2(o(n^{-2})) + n^3(\sqrt{3}/2)^n + \sum_{q=4}^{n} n^q n^{-q-2} < 1 .$$

Q.E.D.

We conjecture that there exists a constant C so that for all n there exists a graph H on n points with $A(H) > n/2 - C$. The following paragraphs are an expository discussion of a probabilistic proof "in progress".*

Let n be a prime, $n \equiv 1$ (modulo 4). Let G be the graph with vertices the integers modulo n and $\{i, j\} \in G$ iff $i - j$ is a square modulo n. Let $\mathbf{G} = \mathbf{G}_{n,p}$ where $p = 10\,000 \log n/n$. Set $\mathbf{H} = G \Delta \mathbf{G}$. Then

$$(14.19) \qquad \sigma(\mathbf{H}) \Delta \mathbf{H} = (\sigma(G) \Delta G) \Delta (\sigma(\mathbf{G}) \Delta \mathbf{G}) .$$

Case 1: $q(\sigma) \geq n/50$. We may find $e_1, \ldots, e_t$, $t > n^2/1000$, such that all e_i, $\sigma^{-1}(e_i)$ are distinct. For $1 \leq i \leq t$ set

$$(14.20) \qquad \mathbf{Y}_i = \begin{cases} 1 & \text{if } e_i \in \sigma(\mathbf{H}) \Delta \mathbf{H} \\ 0 & \text{if } e_i \notin \sigma(\mathbf{H}) \Delta \mathbf{H} . \end{cases}$$

Then the $\mathbf{Y}_i$ are independent 0–1 variables with

$$(14.21) \qquad \text{Prob}\,[\mathbf{Y}_i = 1] = \begin{cases} 2p - p^2 & \text{if } e_i \in \sigma(G) \Delta G \\ (1 - p)^2 & \text{if } e_i \notin \sigma(G) \Delta G \end{cases}$$

and so

$$(14.22) \qquad \begin{aligned} \text{Prob}\,(|\sigma(\mathbf{H}) \Delta \mathbf{H}| < n) &< \text{Prob}\left[\sum_{i=1}^{t} \mathbf{Y}_i < n\right] \\ &< n!^{-1} . \end{aligned}$$

Case 2: $3 \leq q(\sigma) < n/50$. Here, and in Case 3, we use the fact (T. STORER [1967]) that for all $i \neq j$

$$(14.23) \quad |\{k \neq i, j : \{i, k\} \in G \text{ iff } \{j, k\} \notin G\}| = \frac{n-1}{2} .$$

Thus we can show $|G \Delta \sigma(G)| \geq .45\, n\, q(\sigma)$.

$$(14.24) \quad \text{Prob}\left[|\mathbf{H} \Delta \sigma(\mathbf{H})| \leq \frac{n}{2}\right] \leq \text{Prob}\,[|\mathbf{G} \Delta \sigma(\mathbf{G})| \geq n[.45q(\sigma) - .5]]$$

which, since $|\{e : \sigma(e) \neq e\}| < nq(\sigma)$, is very small.

Case 3: $q(\sigma) = 2$. Then $\sigma = (ij)$, $G \Delta \sigma(G) = \frac{n-1}{2}$.

$$(14.25) \qquad \text{Prob}\,[|\mathbf{G} \Delta \sigma(\mathbf{G})| > C(\log n)^2] < n^{-2}$$

for C sufficiently large.

* Substantial improvement on these results, due to the junior author, are available on request.

Combining these cases (see Exercise 2) yields

$$\text{Prob}\left[\exists\, \sigma \neq 1, \mid \mathbf{H} \,\Delta\, \sigma(\mathbf{H}) \mid < \frac{n}{2} - C(\log n)^2\right] < 1 \tag{14.26}$$

so there exists H,

$$A(H) \geq \frac{n}{2} - C(\log n)^2. \tag{14.27}$$

This result can possibly be improved in a number of ways.

(1) Reduce p to $p = 10\,000/n$. Case 2 would remain the same. Case 1 would give Prob $[\mid \sigma(\mathbf{H}) \,\Delta\, \mathbf{H} \mid < n] < e^{-10n}$. If a given σ has $\mid G \,\Delta\, \sigma(G) \mid > n^{1+\delta}$ then Prob $[\mid \sigma(\mathbf{H}) \,\Delta\, \mathbf{H} \mid < n] < p^{n^{1+\delta}} \ll n!^{-1}$. So Case 1 could be handled if it could be shown that $< e^{10n}$ of the σ have $\mid G \,\Delta\, \sigma(G) \mid \leq n^{1+\delta}$. Then Case 3 would give $A(H) > \frac{n}{2} - C \log n$.

(2) Let $\mathbf{G}$ be a random regular graph where each point has degree 10 000. Results on $\mathbf{G}$ similar to those on regular tournaments (Sec. 10) might be required. Case 3 would give $A(H) > \frac{n}{2} - 20\,000$.

(3) For n not a prime $\equiv 1(4)$ we need to find G on $[n]$ so that all points have degree $\frac{n}{2} + \alpha$ with $\mid \alpha \mid < C_1$ and for all i, j

$$\mid \{k : \{i, k\} \in G \text{ iff } \{j, k\} \notin G\} \mid = \frac{n}{2} + \beta \text{ with } \mid \beta \mid < C_2.$$

We turn now to a problem of Gol'dberg. He defines the capacity of a graph as the maximal number of nonisomorphic subgraphs of the graph. He then sets $v(n)$ = the maximal capacity of a graph on n vertices. Clearly $v(n) \leq 2^n$ as any graph has only that many subgraphs. Gol'dberg then asked for $\lim_{n\to\infty} v(n)\, 2^{-n}$.

Gol'dberg's problem appears in V. G. Vizing [1968]. A solution is given by A. D. Koršunov [1971].

Let $\mathbf{G} = \mathbf{G}_{n,1/2}$. We show by very coarse estimates that $\mathbf{G}$ has capacity $> 2^n(1 - o(1))$ with probability approaching 1. This implies $\lim_{n\to\infty} v(n)\, 2^{-n} = 1$. Sharper estimates on $v(n)$ using this method are left as Exercise 5.

Let $S \subseteq \{1, \ldots, n\}$ with $\mid S \mid = t$. We bound the probability that there is some other subgraph isomorphic to $G \mid S$. Let ϱ be a function from the vertices of S to $\{1, \ldots, n\}$. Set $g(\varrho) = \mid \{i : \varrho(i) \neq i\} \mid$. We count the number of edges of the complete graph on S such that $\varrho(e) \neq e$. If $\varrho(i) = i$ and $\varrho(j) \neq j$, $\{i, j\}$ is such an edge. Also, all but at most $g(\varrho)/2$ of the edges $\{i, j\}$ where $\varrho(i) \neq i$, $\varrho(j) \neq j$ are such edges. Setting $g = g(\varrho)$ there are at least

$$g(t-g) + \binom{g}{2} - \frac{g}{2} = g\left(t - \frac{g}{2} - 1\right) \tag{14.28}$$

such edges. We define a graph $H(\varrho)$. Its vertices are the edges e of the complete graph on S and the edges $\varrho(e)$. We connect e to $\varrho(e)$. All points have degree ≤ 2 so the graph consists of isolated points, cycles and paths. We can find $e_1, \ldots, e_s$ so that $e_1, \ldots, e_s$, $\varrho(e_1), \ldots, \varrho(e_s)$ are all distinct, where

$$s \geq g\left(t - \frac{g}{2} - 1\right)\Big/ 3. \tag{14.29}$$

Then Prob $\big(S$ is isomorphic to $\varrho(S)\big) \leq 2^{-s}$. There are less than n^{2g} functions ϱ with $g(\varrho) = g$. Therefore the probability that some $\varrho \neq 1$ gives an isomorphism of $G \mid S$ is bounded by

$$\sum_{g=1}^{t} n^{2g}\, 2^{-g(t-g/2-1)/3} = o(1) \tag{14.30}$$

for $t \geq C \log n$. As almost all $|S| \geq C \log n$,

$$v(n) \geq 2^n\big(1 - o(1)\big). \tag{14.31}$$

Exercises

1. Set $A_1(G) = \min_{\sigma \neq 1} |G \mathbin{\Delta} \sigma(G)|$. Discuss the relationship of $A(G)$ to $A_1(G)$ and prove results analogous to Theorems 14.1 and 14.2 for $A_1(G)$.

2. Give a detailed proof that for $n \equiv 1 \pmod 4$, n prime, there exists H on n vertices

$$A(H) \geq \frac{n}{2} - C(\log n)^2.$$

3. (Thesis Topic) Prove or disprove that there exists a constant C so that for all n there exists H on n vertices

$$A(H) \geq \frac{n}{2} - C.$$

(Disclaimer: The authors make no guarantees that the successful completion of this exercise will be an acceptable dissertation, even if the student's advisor is one of the authors.)

4. In Theorem 14.1 a graph is made symmetric by switching many edges from a single point. It would be interesting to explore notions

of asymmetry where the number of edges switched at a given point would be limited. For example, let $d(D)$ = the maximal degree of a point in the graph D. Set

$$f(n) = \max_{G \subset [n]^2} \min_{\substack{D \subseteq [n]^2 \\ G \triangle D \text{ symmetric}}} d(D) .$$

Prove $f(n) = 0(n)$.

5. Find sharp bounds on the maximal capacity $v(n)$.

6. Prove (14.16). This is actually a very loose bound on $P(\sigma)$.

Problems 7 and 8 deal with variants of Theorem 14.1.

7. Call a graph $G \subseteq [n]^2$ nontrivially symmetric if there exists $\sigma \in S_n$ with no fixed points (i. e. $q(\sigma) = n$) such that $\sigma(G) = G$. Set

$$f(n) = \max_{G \subseteq [n]^2} \min | D | ,$$

where $G \triangle D$ is nontrivially symmetric. Prove $f(n) = \frac{n^2}{6}\left(1 + o(1)\right)$. (Hint: Let G be a random graph for the lower bound. For the upper bound let G be fixed and σ be a random permutation of cycle type $2^{n/2}$).

8. Call a graph $G \subseteq [n]^2$ cyclically symmetric if there exists a cycle $\sigma \in S_n$ such that $\sigma(G) = G$. Set

$$f(n) = \max_{G \subseteq [n]^2} \min | D | ,$$

where $G \triangle D$ is cyclically symmetric. Note $f(n) \leq \frac{1}{2}\binom{n}{2}$ by taking $D = G$ or G^c.

(a) Prove $f(n) = \frac{1}{2}\binom{n}{2}\left(1 + o(1)\right)$.

(b) (?) Give bounds on $\frac{1}{2}\binom{n}{2} - f(n)$.

15. THE IMBALANCE PROBLEMS

Let $A = (a_{ij})$, $a_{ij} = \pm 1$, be an $n \times n$ matrix. (Throughout this section the elements of all matrices are restricted to $\{+1, -1\}$.) We may multiply arbitrary rows and columns by -1 with the object of minimizing the absolute value of the sum of the elements of the matrix. That is, we want to balance the number of $+1$ and -1's as evenly as possible. For example, let $n = 8$ and

$$
\begin{array}{c}
\\
X =
\end{array}
\begin{array}{c}
\\
\left[\begin{array}{cccccccc}
+ & + & + & + & + & + & + & + \\
+ & + & + & - & + & - & - & - \\
+ & - & + & + & - & + & - & - \\
+ & - & - & + & + & - & + & - \\
+ & - & - & - & + & + & - & + \\
+ & + & - & - & - & + & + & - \\
+ & - & + & - & - & - & + & + \\
+ & + & - & + & - & - & + & +
\end{array}\right]
\end{array}
\begin{array}{c}
\text{row sum} \\
8 \\ 0 \\ 0 \\ 0 \\ 0 \\ 0 \\ 0 \\ 2 \\ \hline
\text{sum} = 10
\end{array}
$$

Shifting the second column changes the row sums to 6, -2, 2, 2, 2, -2. 2, 0 with total sum still 10. A row shift changes the sign of the row sum. Shifting the third and fourth rows gives row sums 6, -2, -2, -2, 2, -2, 2, 0 with total sum 2. This is the minimal possible total sum absolute value. For, since n is even, all lines (rows or columns) have even sum $2k$. Therefore, shifting a line changes the total sum by $4k$. Thus, after any number of line shifts the total sum is $\equiv 2$ (modulo 4).

We ask for the minimal S_n so that given any $n \times n$ matrix A we can apply line shifts so that the absolute value of the total sum is $\leq S_n$. More explicitly, set

$$S_n = \max_{(a_{ij})} \min_{x_i,\, y_j} \left| \sum_{i,j=1}^{n} a_{ij} x_i y_j \right| \tag{15.1}$$

where x_i, $y_j = \pm 1$ represent row and column shifts, respectively. If n is even then, setting $A \leq$ an $n \times n$ matrix with sum 2, $S_n \geq 2$. If n is odd then $S_n \geq 1$ since the sum of an odd number of ± 1's can never be zero. The following result was first conjectured by L. Moser.

THEOREM 15.1 (J. KOMLÓS and M. SULYOK [1969]): *There exists n_0 so that for $n \geq n_0$*

$$S_n = \begin{cases} 2 \text{ if } n \text{ is even} \\ 1 \text{ if } n \text{ is odd}. \end{cases}$$

We choose n_0 sufficiently large to justify the approximations we shall make. Fix an $n \times n$ matrix $A = (a_{ij})$. Let ${}_1, \ldots, \vec{\ }_n$ denote the rows of A. If $\vec{b} = (b_1, \ldots, b_n)$, $\vec{c} = (c_1, \ldots, c_n)$ are any two vectors we define

$$\vec{b}\vec{c} = (b_1c_1, \ldots, b_nc_n) \tag{15.2}$$

and

$$s(\vec{b}) = \sum_{i=1}^{n} b_i . \tag{15.3}$$

Let $\vec{\mathbf{y}} = (\mathbf{y}_1, \ldots, \mathbf{y}_n)$ denote a random vector with coordinates ± 1. Then $s(\vec{\mathbf{y}})$ has distribution $\mathbf{S}_n$, as defined in § 3. We note, and this is crucial, that for any fixed $\vec{r}_i$, $\vec{\mathbf{y}}\vec{r}_i$ has the distribution of a random vector with coordinates ± 1 and therefore $s(\vec{\mathbf{y}}\vec{r}_i)$ has distribution $\mathbf{S}_n$.

LEMMA 15.1: *With probability* $> 1 - 1/n$

$$| s(\vec{\mathbf{y}}\vec{r}_i) | < \sqrt{5n \log n} \text{ for } 1 \leq i \leq n . \tag{15.4}$$

PROOF: $\text{Prob}\,[\exists i,\ 1 \leq i \leq n,\ | s(\vec{\mathbf{y}}\vec{r}_i) | \geq \sqrt{5n \log n}]$

$$\leq n \,\text{Prob}\,[| \mathbf{S}_n | \geq \sqrt{5n \log n}] < \frac{1}{n}.$$

We now require an elementary lemma of general interest.

LEMMA 15.2: *If* $\mathbf{X}$ *is a random variable with* $\mathsf{E}(\mathbf{X}) = \mu$, $\text{Max}\,(\mathbf{X}) = N$, *then for* $\lambda \leq \mu$

$$\text{Prob}\,[\mathbf{X} \geq \lambda] \geq (\mu - \lambda)/(N - \lambda). \tag{15.5}$$

PROOF: Let

$$\mathbf{Z} = \begin{cases} 1 \text{ if } X \geq \lambda . \\ 0 \text{ otherwise} . \end{cases}$$

Then

$$\mathbf{Z} \geq (\mathbf{X} - \lambda)/(N - \lambda) \tag{15.6}$$

and

$$\text{Prob}\,(\mathbf{X} \geq \lambda) = \mathsf{E}(\mathbf{Z}) \geq \mathsf{E}\big((\mathbf{X} - \lambda)/(N - \lambda)\big) = (\mu - \lambda)/(N - \lambda). \tag{15.7}$$

LEMMA 15.3: *Set*

$$\mathbf{X} = | \{ i : | s(\vec{\mathbf{y}}\vec{r}_i) | \leq \sqrt{1000 \log n} \} | . \tag{15.8}$$

Then with probability $> 1/n$

$$\mathbf{X} \geq \sqrt{70\, n \log n}\,. \tag{15.9}$$

PROOF:

$$\mathsf{E}(\mathbf{X}) = n \operatorname{Prob}\left[\,|\,\mathbf{S}_n\,| \leq \sqrt{1000 \log n}\,\right] \geq \sqrt{100\, n \log n}$$
(15.10)
$$\operatorname{Max}(\mathbf{X}) = n\,.$$

Lemma 15.3 now follows directly from Lemma 15.2.

Fix $\vec{y}$ satisfying (15.4) and (15.9). Set $M = (a_{ij}y_j) = (m_{ij})$. Denote the rows of M by $\vec{r}_1^*, \ldots, \vec{r}_n^*$ and the row sums by $r_1^*, \ldots, r_n^*$. Set $t = \{\sqrt{1000 \log n}\}$. Then we reorder the rows, for convenience, so that

$$|\,r_i^*\,| \leq r \text{ for } 1 \leq i \leq \sqrt{70n \log n}$$
(15.11)
$$|\,r_i^*\,| \leq \sqrt{5n \log n} \text{ for all } i\,.$$

At this point we deviate from the original proof of J. KOMLÓS and M. SULYOK. Let $\vec{\mathbf{z}} = (\mathbf{z}_1, \ldots, \mathbf{z}_n)$ be a random ± 1 vector with exactly $[(t-1)/2]$ of the $\mathbf{z}_j = -1$.

Case 1 — n odd: If $\vec{r}$ is any ± 1 n-vector with $|\,s(\vec{r})\,| \leq t$ then

$$\operatorname{Prob}\left[\,|\,s(\vec{r}\vec{\mathbf{z}})\,| = 1\right] > 2^{-[(t-1)/2]}\,. \tag{15.12}$$

Hence

$$\mathsf{E}\left[\,|\,\{i :\ 1 \leq i \leq \sqrt{70\, n \log n},\ |\,s(\vec{r}_i^*\vec{\mathbf{z}})\,| = 1\}\,|\,\right]$$
(15.13)
$$> \sqrt{70\, n \log n}\; 2^{-(t-1)/2}\,.$$

We may therefore find $\vec{z}$, set $N = (m_{ij}z_j)$ with row sums $r_1^{**}, \ldots, r_n^{**}$, such that

(1) All $|\,r_i^{**}\,| \leq \sqrt{5n \log n} + 2\,[(t-1)/2]$
(2) $1 \leq |\,r_i^{**}\,| \leq t + 2\,[(t-1)/2]$ for $1 \leq i \leq \sqrt{70\, n \log n}$
(3) $|\,r_i^{**}\,| = 1$ for at least $\sqrt{70n \log n}\; 2^{-(t-1)/2}$ i's.

(In (2), $1 \leq |\,r_i^{**}\,|$ since the sum of an odd number of ± 1 cannot be 0. Since only $[(t-1)/2]$ columns have been shifted, all rows sums change by at most $2[(t-1)/2]$ — which implies upper bounds (1), (2)). Case 1 is completed by Exercise 1.

Case 2 — n even: As in Case 1 we find N with

(1) All $|\,r_i^{**}\,| \leq \sqrt{5n \log n} + 2\,[(t-1)/2]$
(2) $|\,r_i^{**}\,| \leq t + 2\,[(t-1)/2]$ for $1 \leq i \leq \sqrt{70n \log n}$
(3) $|\,r_i^{**}\,| = 2$ for at least $\sqrt{70n \log n}\; 2^{-(t-1)/2}$ i's.

Here there is a slight complication because it may be that many of the $r_i^{**} = 0$ in (2). If less than half of the $r_i^{**} = 0$ in (2) we replac (2) by

$$(2') \qquad 2 \leq |r_i^{**}| \leq t + 2[(t-1)/2] \text{ for } 1 \leq i \leq \frac{1}{2}[\sqrt{70n \log n}]$$

and the proof is completed by Exercise 1. If half of the $r_i^{**} = 0$ in (2) shift *any* one column. Then calling the new rows sums r_i^{***}

(1) All $|r_i^{***}| \leq \sqrt{5n \log n} + 2[(t-1)/2] + 2$

(2) $|r_i^{***}| = 2$ for at least $\frac{1}{2}\sqrt{70n \log n}$ i's.

and, again, the proof is completed by Exercise 1. Q.E.D.

We now wish to use line shifts to maximize the imbalance of $+1$'s and -1's. What is the largest integer B_n so that given any $n \times n$ matrix A we can apply line shifts so that the total sum $\geq B_n$? More explicitly, set

$$(15.14) \qquad B_n = \min_{(a_{ij})} \max_{x_i, y_j} \sum_{i,j=1}^{n} a_{ij} x_i y_j .$$

THEOREM 15.2 (T. A. BROWN and J. H. SPENCER [1971]):

$$\left(\sqrt{\frac{2}{\pi}} + o(1)\right) n^{3/2} < B_n < \left(1 + o(1)\right) n^{3/2}.$$

PROOF: We first show the lower bound. Fix an $n \times n$ matrix X with rows $\vec{r}_1, \ldots, \vec{r}_n$. For any $\vec{y} = (y_1, \ldots, y_n)$ we may choose x_i so that

$$(15.15) \qquad \sum_{j=1}^{n} a_{ij} x_i y_j \geq 0 \text{ for } 1 \leq i \leq n$$

Hence

$$(15.16) \qquad \sum_{i,j=1} a_{ij} x_i y_j = \sum_{i=1}^{n} \left| \sum_{j=1}^{n} a_{ij} y_j \right| = \sum_{i=1}^{n} |s(\vec{y}\vec{r}_i)| .$$

Now let $\vec{\mathbf{y}}$ be random.

$$(15.17) \qquad \mathsf{E}(|s(\vec{\mathbf{y}}\vec{r})|) = \mathsf{E}(|\mathrm{S}_n|) = \sqrt{n}\sqrt{\frac{2}{\pi}}\left(1 + o(1)\right) .$$

Thus

$$(15.18) \qquad \mathsf{E}\left[\sum_{i=1}^{n} |s(\vec{\mathbf{y}}\vec{r}_i)|\right] = n^{3/2}\sqrt{\frac{2}{\pi}}\left(1 + o(1)\right)$$

Therefore we may fix $y_1, \ldots, y_n, x_1, \ldots, x_n$ so that

$$(15.19) \qquad \sum_{i,j=1}^{n} a_{ij} x_i y_j \geq n^{3/2}\sqrt{\frac{2}{\pi}}\left(1 + o(1)\right) .$$

For the lower bound we first consider a slight generalization. Let $B(m, n)$ be the largest integer so that given any $m \times n$ matrix A we can apply line shifts so that the total sum $\geq B(m, n)$. Then $B_n = B(n, n)$. Given m, n there are 2^{mn} possible matrices A. Form equivalence classes by setting $A_1 \sim A_2$ if it is possible to get from A_1 to A_2 by line shifts. It is easy to show that this is an equivalence relation and that there are 2^{m+n-1} matrices in each class. (Note that shifting all rows and all columns gives the original matrix.) Thus there are $2^{mn-m-n+1}$ equivalence classes each of which contains a matrix of total sum $\geq B(m, n)$. Therefore

$$(15.20)\quad \text{Prob}\,[\mathbf{S}_{mn} \geq B(m,n)] = |\{A : \sum_{i,} a_{ij} \geq B(m,n)\}|/2^{mn} \geq 2^{1-m-n}.$$

Applying (3.2),

$$(15.21)\qquad B(m, n) < 2\sqrt{\log 2}\,\sqrt{mn(m+n-1)}\,.$$

In particular,

$$(15.22)\qquad B_n = B(n, n) < 2^{3/2}\sqrt{\log 2}\; n^{3/2}.$$

To improve (15.22) we use a special matrix. An $n \times n$ matrix $H = (h_{ij})$ is called a Hadamard matrix if the n rows are mutually orthogonal and the n columns are mutually orthogonal. Hadamard matrices have been the object of much study. (See, e. g., M. HALL [1967].) For all $\delta > 0$ if n is sufficiently large there exists a Hadamard matrix of some order m, $n(1 - \delta) \leq m \leq n$. (This follows from the existence of Hadamard matrices of order $4^i 12^j$ and a number theory result that these numbers are dense in the above sense.)

Suppose there is a Hadamard matrix of order n and fix such a matrix H. Line shifts preserve the Hadamard property. We show* $B_n \leq n^{3/2}$ by showing that any Hadamard $H = (h_{ij})$ has $\sum_{i,j=1}^{n} h_{ij} \leq n^{3/2}$. Let $\vec{r}_1, \ldots, \vec{r}_n$ denote the rows of H.

For $\vec{v} = (v_1, \ldots, v_n)$ let $\|\vec{v}\| = (\sum_{i=1}^{n} v_i^2)^{1/2}$, the Euclidean Norm. Set $\vec{1} = (1, \ldots, 1)$. Then

$$(15.23)\qquad \sum_{i,j} h_{ij} = (\sum_{i=1}^{n} \vec{r}_i)\cdot\vec{1} \leq \|\sum_{i=1}^{n} \vec{r}_i\|\;\|\vec{1}\|$$

by the Cauchy–Schwartz inequality. Since H is Hadamard, the $\vec{r}_i$ are orthogonal, and

$$(15.24)\qquad \|\sum_{i=1}^{n} \vec{r}_i\| = [\sum_{i=1}^{n} \|\vec{r}_i\|^2]^{1/2} = n\,.$$

Hence

$$(15.25)\qquad \sum_{i,j} h_{ij} \leq \|\sum_{j=1}^{n} \vec{r}_i\|\;\|\vec{1}\| = n^{3/2}.$$

* This argument is due to J. H. LINDSEY.

Now suppose n is not the order of a Hadamard matrix. Set $n = m +$ $+ s$ where a Hadamard H_m exists, $s = o(n)$.

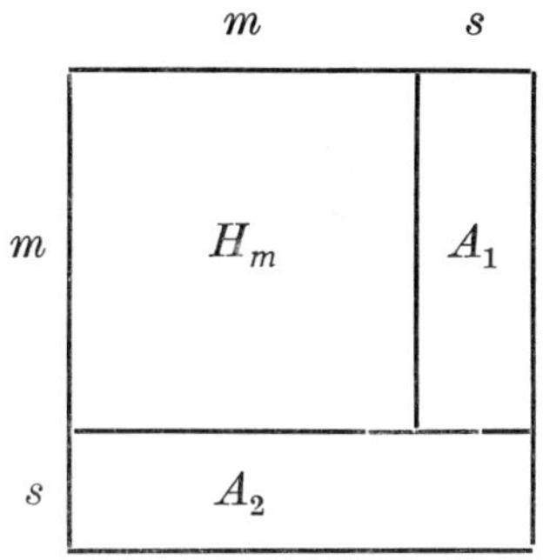

We define an $n \times n$ matrix by its three components in the above figure. H_m is Hadamard. A_1 is such that under any line shifts it has sum $\leq B(m, s)$. A_2 is such that under any line shifts it has sum $\leq B(s, n)$. Then under any line shifts the total matrix has sum $\leq m^{3/2} + B(m, s) + B(n, s) \leq (1 + o(1))n^{3/2}$. Q.E.D.

Y. GORDON and H. S. WITSENHAUSEN [1972] have derived extensions of this result.

Exercises

1. Let $t_1, \ldots, t_n$ be integers such that

(1) All $|t_i| \leq a_1$.
(2) $1 \leq |t_i| \leq a_2$ for at least A_2 i's.
(3) $1 \leq |t_i| \leq a_3$ for at least A_3 i's.

Assume $A_3 \geq a_2$, $A_2 \geq a_1$. Prove there exist $x_1, \ldots, x_n = \pm 1$ such that

$$0 \leq \sum_{i=1}^{n} x_i a_i \leq a_3 .$$

Use this result to complete the proof of Theorem 15.1.

2. (J. SPENCER [unpublished]) Use the proof of Theorem 15.2 to prove Theorem 15.2 for *all* $n > 0$.

(This problem probably requires a small computer. Use Lemmas 15.1–15.3 to find $f_n(k, l)$ such that an $n \times n$ matrix A can be switched so all $|r_i| \leq k$ and $f_n(k, l)$ of the $|r_i| \leq l$. It is probably best to then switch somewhat more that $[(l - 1)/2]$ columns, in order to get many $|r_i| = 1$ or 2.)

3. (a) Generalize (15.18) to prove

$$B(n, m) \geq mE[|\mathbf{S}_n|].$$

(b) Prove that if $2^{n-1} \mid m$ equality holds in (a). That is,

$$B(n, 2^{n-1}K) = K \sum_{i=0}^{(n-1)/2} \binom{n}{i} (n-2i) \text{ if } n \text{ odd}$$

$$K \sum_{i=0}^{n/2} \binom{n}{i} (n-2i) \text{ if } n \text{ even}.$$

(c) Let n be fixed. Prove that for m sufficiently large

$$B(n, m + 2^{n-1}K) = B(n, m) + KB(n, 2^{n-1}) \; (K \geq 0, \text{integral})$$

Hence, $B(n, m)$ may be described by a finite number of "small values" of m and 2^{n-1} linear formulae.

(d) Find $B(2, m)$, $B(3, m)$, $B(4, m)$ for all m.

16. THE EVOLUTION OF RANDOM GRAPHS

In previous sections we have seen how random objects (graphs' matrices, colorings, etc.) can be used to prove combinatorial theorems· In this section we study the properties of random objects directly. We shall require somewhat more knowledge of probability theory then in previous sections. Proofs will be either sketchy or nonexistent and much material will be left as exercises. Most of the results in this section are due to P. ERDŐS and A. RÉNYI.

An important tool in the study of random objects is the "second-moment method". We give an example from Sec. 5. Let $\mathbf{C} = (\mathbf{K}_1, \mathbf{K}_2)$ be a random coloring of $[n]^2$. We showed that $\mathbf{C}$ does not contain a complete monochromatic subgraph on $(2 \log_2 n)(1 + \varepsilon)$ points. (We say $\mathbf{C}$ (or $\mathbf{G}$) has a property if $\mathbf{C}$ (or $\mathbf{G}$) has the property with probability $1 - o(1)$.) We show that this result is best possible in the following sense.

THEOREM 16.1: *Fix* $\varepsilon > 0$. *Let* $t = [(2 \log_2 n)(1 - \varepsilon)]$. *Then* $\mathbf{C}$ *contains a complete monochromatic subgraph on* t *points.*

PROOF: Let S_i, $1 \leq i \leq \binom{n}{t}$, range over the members of $[n]^t$. We show the stronger result that some $[S_i]^2 \subseteq \mathbf{K}_1$. Set

$$X_i = \begin{cases} 1 & \text{if } [S_i]^2 \subseteq \mathbf{K}_1 \\ 0 & \text{otherwise}. \end{cases} \tag{16.1}$$

$$\mathbf{Y} = \sum_{i=1}^{\binom{n}{t}} \mathbf{X}_i \tag{16.2}$$

Set

$$\mu = 2^{-\binom{t}{2}} = \mathsf{E}(\mathbf{X}_i). \tag{16.3}$$

Then

$$\mathsf{E}(\mathbf{Y}) = \binom{n}{t} \mu \gg 1. \tag{16.4}$$

In Sec. 5, with $t = (2 \log_2 n)(1 + \varepsilon)$, $\mathsf{E}(\mathbf{Y}) \ll 1$ and Prob $[\mathbf{Y} > 0] \leq \leq \mathsf{E}(\mathbf{Y}) \ll 1$ trivially follows. Given (16.4) it does not automatically follow that

$$\text{Prob}\,[\mathbf{Y} = 0] = o(1)\ (?) \tag{16.5}$$

(See Exercise 8.) We must study the second moment $\text{Var}(\mathbf{Y})$. In general

$$\text{Prob}\,(\mathbf{Y} = 0) \leq \text{Var}\,(\mathbf{Y})/\mathsf{E}(\mathbf{Y})^2 . \tag{16.6}$$

Therefore to prove (16.5), and thus Theorem 16.1, we need to show

$$\text{Var}\,(\mathbf{Y}) = o(\mathsf{E}(\mathbf{Y})^2)\ (?) \tag{16.7}$$

$$\begin{aligned} \text{Var}\,(\mathbf{Y}) &= \mathsf{E}(\mathbf{Y}^2) - \mathsf{E}(\mathbf{Y})^2 \\ &= \left[\sum_{i,j=1}^{\binom{n}{t}} \mathsf{E}(\mathbf{X}_i\mathbf{X}_j)\right] - \left[\binom{n}{t}\mu\right]^2 \\ &= \sum_{i,j=1}^{\binom{n}{t}} [\mathsf{E}[\mathbf{X}_i\mathbf{X}_j] - \mu^2] \end{aligned} \tag{16.8}$$

$$\begin{aligned} \mathsf{E}(\mathbf{X}_i\mathbf{X}_j) &= \text{Prob}\,[\mathbf{X}_i = 1 = \mathbf{X}_j] \\ &= \text{Prob}\,[\mathbf{X}_i = 1 \mid \mathbf{X}_j = 1]\,\text{Prob}\,[\mathbf{X}_j = 1] \\ &= \text{Prob}\,[\mathbf{X}_i = 1 \mid \mathbf{X}_j = 1]\,\mu . \end{aligned} \tag{16.9}$$

Hence

$$\begin{aligned} \text{Var}\,(\mathbf{Y}) &= \binom{n}{t}^2 \mathsf{E}_{\mathbf{ij}}[\mathsf{E}[\mathbf{X}_\mathbf{i}\mathbf{X}_\mathbf{j}] - \mu^2] \\ &= \binom{n}{t}^2 \mu^2 \mathsf{E}\left[\frac{\text{Prob}\,(\mathbf{X}_\mathbf{i} = 1 \mid \mathbf{X}_\mathbf{j} = 1)}{\text{Prob}\,(\mathbf{X}_\mathbf{i} = 1)}\right] \end{aligned} \tag{16.10}$$

where $\mathbf{i}, \mathbf{j}$ are independent random elements of $\left[\binom{n}{t}\right]$. Thus it suffices to show

$$\mathsf{E}\left[\frac{\text{Prob}\,(\mathbf{X}_\mathbf{i} = 1 \mid \mathbf{X}_\mathbf{j} = 1)}{\text{Prob}\,(\mathbf{X}_\mathbf{i} = 1)}\right] = 1 + o(1)\ (?) \tag{16.11}$$

The LHS of (16.11) is $\mathsf{E}\left[2^{\binom{\mathbf{u}}{2}}\right]$ where $\mathbf{u} = |S_\mathbf{i} \cap S_\mathbf{j}|$. By symmetry we may assume $S_\mathbf{j} = S_j =$ a fixed t-set. Then the LHS of (16.11) is

$$\sum_{\nu=0}^{t} \left[\frac{\binom{t}{v}\binom{n-t}{t-v}}{\binom{n}{t}}\right] 2^{\binom{\nu}{2}} = 1 + o(1) \tag{16.12}$$

by an elementary computation. Q.E.D.

In general, if $\mathbf{Y}$ is the sum of identically distributed 0–1 variables $\mathbf{X}_i$ and $E(\mathbf{Y}) \to \infty$, (16.11) implies (16.5). (See Exercises 1, 2.)

Let $\mathbf{G} = \mathbf{G}_{n,e}$ be a random graph on n points, e edges. We examine the evolution of $\mathbf{G}$ as e increases. Unreferenced results are due to P. Erdős and A. Rényi (1960). Let A be a property of some graphs (e. g.: connectivity, planarity). Set

$$P_{n,e}(A) = \text{Prob}\,[\mathbf{G}_{n,e} \text{ has property } A]. \tag{16.13}$$

$A(n)$ is called a threshold function for A if

$$\lim_{n\to\infty} P_{n,N(n)}(A) = \begin{cases} 0 \text{ if } N(n)/A(n) \to 0 \\ 1 \text{ if } N(n)/A(n) \to \infty. \end{cases} \tag{16.14}$$

In some cases there is a distribution function $G(x)$ and functions $A_1(n)$, $A_2(n)$ so that

$$\operatorname*{Lim}_{n\to\infty} P_{n,N(n)}(A) = G(x) \text{ if } \lim_{n\to\infty} \frac{N(n) - A_1(n)}{A_2(n)} = x\,. \tag{16.15}$$

If G has n vertices, e edges, we say G has degree $\frac{2e}{n}$. We call G *balanced* if all subgraphs H of G have degree $(H) \leq$ degree (G). In particular, trees, cycles, complete graphs, and complete bipartite graphs are balanced.

If $e = o(\sqrt{n})$, $\mathbf{G} = \mathbf{G}_{n,e}$ consists of disjoint edges. If $e = c\sqrt{n}$ then trees on three points appear with positive probability. The general result for the appearance of small subgraphs is:

Theorem 16.2 (P. Erdős and A. Rényi [1960]): *Let H be a connected balanced graph on k points with l edges. The threshold function for the property of containing a subgraph isomorphic to H is $n^{2-k/l}$.* The proof is left as Exercise 4. In particular, trees on k points appear at $e = n^{(k-2)/(k-1)}$. Complete bipartite $a \times a$ graphs appear at $e = n^{2-2/a}$. (Compare with Sec. 13.)

Now assume $e = cn$ where c is fixed. The nature of $\mathbf{G}_{n,e}$ changes dramatically at $c = 1/2$. If $c < 1/2$ the largest component of $\mathbf{G}$ has size $(1 + o(1)) \log n/(2c - 1 - \log 2c)$. (See Exercise 5.) If $c = \frac{1}{2}$ the largest component has order of magnitude $n^{2/3}$. If $c > \frac{1}{2}$ the largest component has size $h(c)n(1 + o(1))$ where h is a specific continuous function, $h\left(\frac{1}{2}\right) = 0$, $h(1) = 1$. This "double jump" is one of the most surprising results in the theory.

Planarity also changes at $c = \frac{1}{2}$. If $c < \frac{1}{2}$, $\mathbf{G}$ is planar (Exercise 6). If $c > \frac{1}{2}$, $\mathbf{G}$ is not planar (Exercise 7). Even stronger, if $e = \frac{n}{2} + \sqrt{n}\,\omega(n)$ where $\omega(n) \to \infty$ then $\mathbf{G}$ is not planar. If $e = \frac{n}{2} + \lambda\sqrt{n}$, Prob [$\mathbf{G}$ is planar] $= g(\lambda) > 0$. The function g is not known (except that $g(\lambda) > 0$ for $-\infty < \lambda$ and $\lim_{\lambda \to \infty} g(\lambda) = +1$). It is even possible that $g(\lambda) = 1$ for all λ, though this seems doubtful. If $\lim_{\lambda \to -\infty} g(\lambda) = 0$ then we would have a distribution function (16.15). The results for $e = \frac{n}{2} + \lambda \sqrt{n}$ are so delicate that they may not apply to $\mathbf{G} = \mathbf{G}_{n,p}$, $p = e \Big/ \binom{n}{2}$.

The chromatic number $\chi(\mathbf{G})$ also changes at $c = \frac{1}{2}$. If $c < \frac{1}{2}$ then Prob [$\mathbf{G}$ is a forest] $= p(c) > 0$ so Prob $[\chi(\mathbf{G}) = 2] > 0$. If $c = \frac{1}{2}$, $\chi(\mathbf{G}) \geq 3$. Now assume c is large, but fixed. Then

$$(16.16) \qquad \frac{c}{2 \log c}(1 + o(1)) \leq \chi(\mathbf{G}) \leq \frac{c}{\log c}(1 + o(1)) .$$

In (16.16) the $o(1)$ notation is with respect to c. (See Exercise 9.)

We now let $e \sim \frac{1}{2} n \log n$. At this point the last isolated points disappear and the graph becomes connected. There is a surprisingly precise result. Let A be connectivity. Then (16.15) holds with $A_1(n) = \frac{1}{2} n \log n$, $A_2(n) = n$, $G(x) = e^{-e^{-2x}}$. More intuitively, a random graph on n points, $\frac{1}{2} n \log n + xn$ edges, x fixed is connected with probability $e^{-e^{-2x}}$. (Exercise 13.)

Now let the number of vertices $= n$ be even (trivial modifications are required with n odd). A 1-factor in a graph G is a set of $\frac{n}{2}$ vertex disjoint edges. P. Erdős and A. Rényi [1966] proved that if $e = \frac{1}{2} n \log n + \omega(n) \to \infty$ then $\mathbf{G}$ has a 1-factor (see Exercise 10, 11).

A graph is called Hamiltonian if it contains a cycle passing through every point exactly once. Note that if G is Hamiltonian it contains a 1-factor. One of the most interesting unsolved problems in this subject is to find a threshold function for Hamiltonality. Almost certainly $e = n^{1+\varepsilon}$, $\varepsilon > 0$, suffices and we conjecture that $n \log n$ is the threshold function (see Exercise 12).

Exercises

1. If C is a 2-coloring of $[n]^2$ set $k(C) = \max |S|$, $[S]^2 \subseteq K_i$ for $i = 1$ or 2. Set $t =$ minimal integer, $\binom{n}{t} 2^{1-\binom{t}{2}} > 1$. Prove that

$$|k(\mathbf{C}) - t| \leq 2 .$$

2. If T is a tournament set $v(T) = \max t$, $\exists\, A, B$, disjoint, $|A| = |B| = t$, $A \times B \subseteq T$. Set $t =$ minimal integer, $\binom{n}{2t}\binom{2t}{t}2^{-\binom{t}{2}} > 1$. Prove $|v(\mathbf{T}) - t| \le 2$.

3. If σ is a permutation of $[n]$ set $v(\sigma) =$ the maximum length of a monotonic subsequence of the sequence $\sigma(1), \ldots, \sigma(n)$.

P. Erdős and G. Szekeres (1935) proved $v(\sigma) \ge \{\sqrt{n}\}$ for *all* σ. S. Ulam conjectures

$$(?) \qquad v(\boldsymbol{\sigma}) = \big(2 + o(1)\big)\sqrt{n}\,.$$

Note: In the original Erdős, Rényi papers many of the following exercises were done for $\mathbf{G} = \mathbf{G}_{n,e}$, a random graph with n points, e edges. We instead look at $\mathbf{G}_{n,p}$ where $p \sim e\Big/\binom{n}{2}$.

4. Let $\mathbf{G} = \mathbf{G}_{n,p}$. Let H be a connected balanced graph of k points, l edges.

(a) If $p = o(n^{-k/l})$ prove that $\mathbf{G}$ does not contain a subgraph isomorphic to H. (H need not be balanced.)

(b) If $p/n^{-k/l} \to \infty$ prove that $\mathbf{G}$ does contain a subgraph isomorphic to H. [Hint: Let S_i range over $[n]^k$, $1 \le i \le \binom{n}{k}$. Set $\mathbf{X}_i = 1$ if $\mathbf{G} \mid S_i$ contains a subgraph isomorphic to H, $\mathbf{X}_i = 0$ otherwise. Show (16.11).]

5. Consider the following Galton–Watson process. We begin with one (active) member. In each generation, each active member "spawns" $\mathbf{X}$ (active) members and then becomes inert. Let $\mathbf{X}$ be Poisson of mean $\lambda < 1$. The numbers spawned are assumed to be independent. Let $P_r =$ the probability that when the process stops there are exactly r members. R. Otter (1949) proved

$$P_r = \frac{(r\lambda)^{r-1}}{r\,!}\, e^{-r\lambda}.$$

Let $\mathbf{G} = \mathbf{G}_{n,p}$, $p = 2c/n$, $c < \dfrac{1}{2}$ fixed. Use the above result to prove that the largest component of G has size

$$\big(1 + o(1)\big) \log n/(2c - 1 - \log 2c)\,.$$

6. Let $\mathbf{G} = \mathbf{G}_{n,p}$, $p = 2\,c/n$, $c < \dfrac{1}{2}$. Let $\omega(n)$ be a function slowly approaching infinity [e. g.: $\omega(n) = \log\log n$].

(a) Prove that fewer than $\omega(n)$ points of $\mathbf{G}$ are in cycles.

(b) Prove no components of size $\le \omega(n)$ have more edges than points.

(c) Use (a), (b) to prove that $\mathbf{G}$ is planar.

7. Let $\mathbf{G} = \mathbf{G}_{n,p}$, $p = 2c/n$, $c > \frac{1}{2}$. Set $k = \left[\frac{10 \log n}{\log (2c)}\right]$. Let $\mathbf{X}$ be the number of k-cycles in $\mathbf{G}$ that contain points P, Q, R, P', Q', R' — in order around the cycle — such that $(P, P'), (Q, Q'), (R, R') \in \mathbf{G}$.

(a) Prove $\mathsf{E}(\mathbf{X}) \to \infty$.

(b) Prove Prob $[\mathbf{X} > 0] \to 1$ by using (16.11).

(c) Use (a), (b) to prove that $\mathbf{G}$ is not planar.

8. Set $p = n^{-1}$, $\mathbf{G} = \mathbf{G}_{n,p}$. Set $t(\mathbf{G})$ = the number of maximal trees on $n - 5$ points.

(a) Prove $\mathsf{E}(t(\mathbf{G})) = \binom{n}{5} (n-5)^{n-7} p^{n-6}(1-p)^{5(n-5)}$. (Hint: The number of labeled trees on m points is m^{m-2}.)

(b) Prove Prob $[t(\mathbf{G}) > 0] < \binom{n}{5} (1 - e^{-1})^{n-5} = o(1)$.

9. Let $p = 2c/n$, c fixed, large, $\mathbf{G} = \mathbf{G}_{n,p}$.

(a) Prove $\chi(\mathbf{G}) \geq \frac{c}{\log c}(1 + o(1))$ where $o(1)$ is with respect to c. (Hint: Use Sec. 11, Method 1.)

(b) Prove $\chi(\mathbf{G}) \leq \frac{2c}{\log c}(1 + o(1))$. [This requires a new technique. Pick $\mathbf{P}_1, \mathbf{P}_2, \ldots$ from $[n]$ where $\mathbf{P}_i$ is picked randomly from $\{P : (P, P_j) \notin \mathbf{G},\ P \neq P_j,\ 1 \leq j < i\}$. Show this yields an independent set $\mathbf{K}_1 = \{\mathbf{P}_1, \ldots, \mathbf{P}_a\}$, $a = n\left[\frac{2c}{\log 2c}\right](1 - o(1))$ with probability $1 - o(n^{-1})$. Note $\mathbf{G} - \mathbf{K}_1 \sim \mathbf{G}_{n-a,p}$ and use induction on c (over all real c).]

(c) (?) Let $c = c(n)$ be a function of n. Assume $c(n) \to \infty$, $c(n) = o(n)$. Prove

$$\frac{c}{\log c}[1 + o(1)] \leq \chi(G) \leq \frac{2c}{\log c}[1 + o(1)]$$

where $o(1)$ is with respect to n.

10. (P. Erdős and A. Rényi [1963]) Let $\mathbf{A} = \mathbf{A}_{n,p}, = (\mathbf{a}_{ij})$ be a random $n \times n$ 0–1 matrix with Prob $[\mathbf{a}_{ij} = 1] = p$. Let $p = \frac{\log n}{n} + \frac{c}{n}$. Prove

$$\text{Prob}\,[\text{Perm}\,(\mathbf{A}) > 0] = e^{-2e^{-c}}.$$

Hint: Perm $(A) = 0$ iff $\exists\, k \times (n - k + 1)$ all 0 submatrix (P. König [1950] or see, e. g., M. Hall [1967, Thm. 5.1.4]). Show Prob $[\exists\, k \times (n - k + 1)$ all 0 submatrix, $k \neq 1, n$, that cannot be extended to 0 line $= o(1)$.]

11. (P. ERDŐS and A. RÉNYI [1966]). Let $\mathbf{G} = \mathbf{G}_{n,p}$, $p = \frac{\log n}{n} + \frac{\omega(n)}{n}$, $\omega(n) \to \infty$. Prove that $\mathbf{G}$ contains a 1-factor. [Hint: W. TUTTE (1947) proved G has a 1-factor iff $G - \{P_1, \ldots, P_r\}$ has $\leqq r + 1$ connected components with an odd number of points. The proof of the exercise is lengthy, even with TUTTE's result.]

12. V. CHVATAL [private correspondence] defines a graph G to be t-compact if for all k the deletion of $\leqq tk$ points leaves $\leqq k$ components. He conjectures that if G is 2-compact it is Hamiltonian.

Set $\mathbf{G} = \mathbf{G}_{n,p}$, $p = c \log n/n$. Prove that for every t there exists $c = c(t)$ so that $\mathbf{G}$ is t-compact.

13. Let $\mathbf{G} = \mathbf{G}_{n,p}$, $p = \frac{\log n}{n} + \frac{2x}{n}$, x fixed. Prove

$$\text{Prob}\,[\mathbf{G} \text{ is connected}] = e^{-e^{-2x}}.$$

(Hint: First show $\mathbf{G}$ consists of a connected component and isolated points.)

14. Let $c(G)$ = the number of cliques in G. Set $\mathbf{G} = \mathbf{G}_{n,1/2}$. Find $\mathsf{E}[c(\mathbf{G})]$, Var $[c(\mathbf{G})]$ and prove Var $[c(\mathbf{G})] = o\big[\mathsf{E}[c(\mathbf{G})]\big]$.

17. THE KITCHEN SINK

In this final section we present a potpourri of problems and results that did not fit into the preceding sections. Proofs shall only be out lined.

A. Infinite Random Graphs

Let N be the set of positive integers. We wish to put a measure μ on the set of all graphs $G \subseteq [N]^2$. Intuitively, we wish the graph to be obtained by flipping a fair coin for each $e \in [N]^2$ to determine if $e \in G$. That is,

$$\mu[\{G : e \in G\}] = 1/2$$

and the events $e \in G$ are independent over different $e \in [N]^2$. We induce the canonical Lebesque measure generated by these conditions. We may then speak of a random graph $\mathbf{G} \subseteq [N]^2$. The theory of infinite random graphs is demolished by the following:

THEOREM 17A (P. ERDŐS and A. RÉNYI [1963b]): *There exists* $G_0 \subseteq [N]^2$ *so that*

$$\text{Prob}\,[G_0 \cong \mathbf{G}] = 1\,.$$

That is, the infinite random graph always looks the same. To prove this we first say that a graph $G \subseteq [N]^2$ has property U (for universal) if for all distinct $u_1, \ldots, u_n, v_1, \ldots, v_m \in N$ there exists $w \in N$, different from the u's and v's, so that $\{w, u_i\} \in G$, $1 \leq i \leq n$, and $\{w, v_j\} \notin G$, $1 \leq j \leq m$.

LEMMA: *If* G_1, G_2 *have property* U *then* $G_1 \cong G_2$.

PROOF: We define $\sigma : N \to N$ by determining $\sigma(1)$, $\sigma^{-1}(1)$, $\sigma(2)$, $\sigma^{-1}(2), \ldots$ in order. When $\sigma(i)$ need to be determined σ has been determined on a finite set V. By property U we find, and fix, $\sigma(i)$ so that

$$\{i, j\} \in G_1 \text{ iff } \{\sigma(i), \sigma(j)\} \in G_2, \quad j \in V\,.$$

We define $\sigma^{-1}(i)$ analogously. Of course, if for some j $\sigma^{-1}(j)$ had been determined as i then $\sigma(i)$ is set at j—similarly for $\sigma^{-1}(i)$. After the countable procedure σ is a well-defined isomorphism of G_1 onto G_2.

PROOF of Theorem: For any $m, n, u_1, \ldots, u_m, v_1, \ldots, v_n$

$$\mu[\{G : \nexists w, \{w, u_i\} \in G, \{w, v_j\} \notin G\}] \leq (1 - 2^{-(m+u)})^K$$

for all K and consequently the measure is zero. There are only a countable number of $m, n, u_1, \ldots, u_m, v_1, \ldots, v_n$. The union of a countable collection of sets of measure zero has measure zero. Hence

$$\mu[\{G : G \text{ does not have property U}\}] = 0$$

so, fixing a G_0 with property U,

$$\text{Prob}\,[G_0 \cong \mathbf{G}] = \mu[\{G : G \text{ has property U}\}] = 1$$

Q.E.D.

B. A \$ 300 Problem

P. ERDŐS and L. MOSER posed the following question: What is the maximal number of integers $0 < a_1 < \ldots < a_k \leq n$ such that all $\sum_{i \in S} a_i$, $S \subseteq [k]$ are distinct. (The empty sum is considered zero for convenience.) Denote by $g(n)$ this maximal value of k. By setting $a_i = 2^{i-1}$ we see

$$g(n) \geq 1 + [\log_2 n].$$

Since there are $2^{g(n)}$ sums, all between 0 and $ng(n)$,

$$2^{g(n)} \leq ng(n)$$

so

$$g(n) \leq \log_2 n + \log_2 \log_2 n + o(1). \quad (*)$$

It is not known whether

$$g(n) = \log_2 n + o(1). \quad (?)$$

P. ERDŐS has offered \$300 for a proof or disproof of (?). At this writing the U.S. dollar is under severe stress.* Therefore, the prize may be changed into an equivalent amount of gold in the near future.

R. K. GUY and J. CONWAY [unpublished] have shown, by constructive means, that $g(2^m) \geq m + 2$ for $m \geq 23$.

An improvement of (*) is given by the following result of P. ERDŐS and L. MOSER (see P. ERDŐS [1955]):

THEOREM 17B: $g(n) \leq \log_2 n + \dfrac{1}{2} \log_2 \log_2 n + o(1)$.

The proof involves the "second-moment method". Let $0 < x_1 < \ldots < x_k \leq n$ be such that $\sum_{i \in S} x_i$ are distinct. Let $\mathbf{s} = \sum_{i \in \mathbf{s}} x_i$,

* These words were originally written in 1970. Three years later they remain equally valid!

where $\mathbf{S}$ is a random subset of $[k]$. Then

$$\mathrm{E}(\mathbf{s}) = \frac{1}{2} \sum_{i=1}^{k} x_i = \mu$$

$$\mathrm{Var}\,(\mathbf{s}) = \sum x_i^2/4 \leq kn^2/4\,.$$

Thus

$$\mathrm{Prob}\,[|\,\mathbf{s} - \mu\,| < \sigma] > 1 - \mathrm{Var}\,(\mathbf{s})/\sigma^2 \geq 1 - kn^2/4\sigma^2\,.$$

But

$$\mathrm{Prob}\,[|\,\mathbf{s} - \mu\,| < \sigma] = |\,\{A : |\sum_{i \in A} x_i - \mu\,| < \sigma\}\,|\,2^{-k} \leq 2\sigma\,2^{-k}$$

since, by assumption, all $\sum_{i \in A} x_i$ are distinct. Consequently,

$$2\sigma\,2^{-k} \geq 1 - kn^2/4\sigma^2\,.$$

The theorem follows by setting $\sigma = n\sqrt{k}$.

C. A Coin-Weighing Problem

A collection $\{S_1, \ldots, S_m\}$ of subsets of $[n]$ is called a determining collection if an arbitrary subset T of $[n]$ is uniquely determined by the cardinalities $|\,S_i \cap T\,|$, $1 \leq i \leq m$. We now study the minimum value $D(n)$ of m for which a determining collection exists.

The problem is often expressed as a coin-weighing problem. Let n coins of weights 0 and 1 be given. We are also given a scale with which we may weigh any subset of the coins. The object is to determine the weights of the coins with the minimal number of weighings. The weighings must be predetermined—i. e., information gained on previous weighings cannot be used in later weighings.

By weighing each coin ($S_i = \{i\}$) we see $D(n) \leq n$. The asymptotic result, however, is given by

Theorem 17C: $D(n) = n \log(4)/\log(n) + O(n \log\log(n)/\log^2(n))$.

For each i there are only $n + 1$ possible $|\,S_i \cap T\,|$. Therefore, by the pigeon-hole principle, if $\{S_1, \ldots, S_m\}$ is a determining collection $(n+1)^m > 2^n$. This gives the bound $D(n) > n \log 2/\log(n+1)$. Note that this bound holds even if information from previous weighings may be used.

We can improve the lower bound by using the second-moment method. This proof is due to L. Moser (see B. Lindström [1965] pp. 488f). Let $\{S_1, \ldots, S_m\}$ be a determining collection with $m = D(n)$. Let A be the $n \times m$ incidence matrix of the S_i (i. e., $a_{ij} = 1$ iff $i \in S_j$). We assume, for convenience, that the coins weigh -1 and $+1$. Then $\vec{x} = (x_1, \ldots, x_n)$, $x_i = \pm 1$ is determined uniquely by $\vec{x}A$. Let $\vec{\mathbf{x}} = (\mathbf{x}_1, \ldots, \mathbf{x}_n)$ be a random n-vector, $\mathbf{x}_i = \pm 1$. Let $||\cdot||$ denote the usual Euclidean Norm in R^m. Set $\vec{\mathbf{y}} = (\mathbf{y}_1, \ldots, \mathbf{y}_m) = \vec{\mathbf{x}}A$

$$\mathsf{E}[\|\vec{\mathbf{y}}\|^2] = \mathsf{E}[\sum_{i=1}^{m} \mathbf{y}_i^2] = \sum_{i=1}^{m} \mathsf{E}[\mathbf{y}_i^2].$$

Whatever A is, $\mathbf{y}_i$ has distribution $\mathbf{S}_n$ so

$$\mathsf{E}[\mathbf{y}_i^2] = \mathsf{E}[\mathbf{S}_n^2] = n$$

and

$$\mathsf{E}[\|\vec{\mathbf{y}}\|^2] = mn.$$

Therefore

$$\text{Prob}\,[\|\vec{\mathbf{y}}\|^2 \leq 2mn] \geq \frac{1}{2}.$$

That is, there are at least 2^{n-1} vectors $\vec{x}$ such that $\|\vec{x}A\|^2 \leq 2mn$. But all $\vec{x}A$ are distinct and have integral coefficients. The number of points $\vec{y} \in R^m$ with integral coefficients, $\|\vec{y}\|^2 \leq 2mn$, is bounded by the volume of the sphere of radius $(2mn)^{1/2}$ in R^m which is $(c_1 n)^{m/2}$. Thus

$$2^{n-1} \leq (c_1 n)^{m/2}$$

so

$$D(n) = m \geq [n \log 4/\log n]\big(1 + o(1)\big).$$

We may obtain a lower bound by probabilistic means. Assume, for convenience, that $S_1 = [n]$. Let $\mathbf{S}_2, \ldots, \mathbf{S}_m$ be random subsets of $[n]$. Let A^*, B^* be distinct sets that are not separated. Then $A = A^* - B^*$, $B = B^* - A^*$ are disjoint sets which are not separated.

$$\mathsf{E}[|\{(A, B) : A \cap B = \emptyset, A, B \text{ not separated}\}|]$$

$$\leq \sum_{t=1}^{n/2} \binom{n}{t}\binom{n-t}{t} \text{Prob}\,[A, B \text{ not separated} \mid |A| = |B| =$$

$$= t, A \cap B = \emptyset] \leq \sum_{t=1}^{n/2} \binom{n}{t}\binom{n-t}{t}\left(\frac{c_2}{\sqrt{t}}\right)^{m-1} = o(1)$$

if $m \geq 2\,n \log 4/\log n$. Thus $D(n) \leq 2\,n \log n/\log 4$.

The upper bound must, however, be considered a victory for the constructivists. D. Cantor and W. Mills [1966] and B. Lindström [1965] proved that

$$D(n) \leq \frac{n \log 4}{\log n} + o(n \log \log n/\log n).$$

Their methods were completely constructive.

This volume ends on a melancholy note—with a problem to which the probabilistic methods did not give the best answer. We wish our readers more complete success.

REFERENCES

ABBOTT, H. L. (1965), An Application of Ramsey's Theorem. To a Problem of Erdős and Hajnal, *Canad. Math. Bull.* **8,** 515–517.

BEINEKE, L. W. and HARARY, F. (1965), The Maximum Number of Strongly Connected Subtournaments. *Canad. Math. Bull.* **8,** 491–498.

BERLEKAMP, E. R. (1968), A Construction For Partitions which Avoid Long Arithmetic Progressions, *Canad. Math. Bull.* **11,** 409–414.

BROWN, T. A. and SPENCER, J. H. (1971), Minimization of ± 1 Matrices Under Line Shifts, *Colloq. Math.* (Poland) **23,** 165–171.

BRUCK, R. H. and RYSER, H. J. (1949), The Nonexistence of Certain Finite Projective Planes, *Canad. J. Math.* **1,** 88–93.

CANTOR, D. G. and MILLS, W. H. (1966), Determination of a Subset from Certain Combinatorial Properties, *Canad. J. Math.* **18,** 42–48.

CHERNOFF, H. (1952), A Measure of Asymptotic Efficiency for Tests of a Hypothesis Based on the Sum of Observations, *Ann. Math. Stat.* **23,** 493–509.

COLOMBO, U. (1964), Sui Circuiti nei Grafi Completi. *Boll. Un. Mat. Ital.* **19,** 153–170.

CHVATAL, V. (1970), Hypergraphs and Ramseyian Theorems, Ph. D. Thesis, University of Waterloo, Waterloo, Ontario, Canada.

CHVATAL, V. (1971), Hypergraphs and Ramseyian Theorems, *Proc. Amer. Math. Soc.* **27,** 434–440.

BLANCHE DESCARTES (see W. Tutte).

DOYEN, J. (1970), Sur La Croissance du Nombre de Systèmes Triples de Steiner non Isomorphes, *J. Comb. Theory* **8,** 424–441.

ENTRINGER, R. C., ERDŐS, P. and HARNER, C. Some External Properties Concerning Transitivity in Graphs, *Periodica Mathematica Hungarica* (to appear).

ERDŐS, P. (see R. C. Entringer).

ERDŐS, P. (1947), Some Remarks on the Theory of Graphs. *Bulletin Amer. Math. Soc.* **53,** 292–294.

ERDŐS, P. (1955), Problems and Results in Additive Number Theory, in Colloq. sur la Théorie des Nombres (Bruxelles) 127–137.

ERDŐS, P. (1959), Graph Theory and Probability, *Canadian J. of Math.* **11,** 34–38.

ERDŐS, P. (1961), Graph Theory and Probability II. *Canad. J. of Math.* **13,** 346–352.

ERDŐS, P. (1962), On Circuits and Subgraphs of Chromatic Graphs, *Mathematika* **9,** 170–175.

ERDŐS, P. (1963), On a Problem in Graph Theory. *Math. Gazette* **47,** 220–223.

ERDŐS, P. (1964), On a Combinatorial Problem II, *Acta Math. Acad. Sci. Hung.* **15,** 445–447.

ERDŐS, P. (1965), On Extremal Problems of Graphs and Generalized Graphs, *Israel J. of Math.* **2,** 189–190.

ERDŐS, P. (1967), Some Remarks on Chromatic Graphs, *Colloquium Mathematicum* **16,** 253–256.

ERDŐS, P. (1969), On a Combinatorial Problem III, *Canad. Math. Bull.* **12,** 413–416.

ERDŐS, P. and HAJNAL, A. (1961), On a Property of Families of Sets, *Acta. Math. Acad. Sci. Hung.* **12,** 87–123.

ERDŐS, P. and HAJNAL, A. (1966), On Chromatic Number of Graphs and Set Systems, *Acta Math. Acad. Sci. Hung.* **17,** (1–2), 61–99 (esp. 94–99).

ERDŐS, P. and HAJNAL, A. (1968), On a Combinatorial Problem (in Hungarian), *Mat. Lapok* **19**, 345–348.

ERDŐS, P. and HANANI, H. (1963), On a Limit Theorem in Combinatorial Analysis, *Publ. Math. Debrecen* **10**, 10–13.

ERDŐS, P. and MOON, J. W. (1964), On Subgraphs of the Complete Bipartite Graph, *Canad. Math. Bull.* **7**, 35–39.

ERDŐS, P. and MOON, J. W. (1965), On Sets of Consistent Arcs in a Tournament. *Canad. Math. Bull.* **8**, 269–271.

ERDŐS, P. and MOSER, L. (1964), A Problem on Tournaments, *Canad. Math. Bull.* **7**, 351–356.

ERDŐS, P. and RADO, R. (1952), Combinatorial Theorems on Classifications of Subsets of a Given Set, *Proc. of the London Math. Soc.*, (3) **2**, 417–439.

ERDŐS, P. and RÉNYI, A. (1960), On the Evolution of Random Graphs, *Mat. Kutató Int. Közl.* **5**, 17–60.

ERDŐS, P. and RÉNYI, A. (1963*a*), On Random Matrices, *Publ. Math. Inst. Hung. Acad. Sci.* **8**, 455–461.

ERDŐS, P. and RÉNYI, A. (1963*b*), Asymmetric Graphs, *Acta Math. Acad. Sci. Hung.* **14**, 295–315.

ERDŐS, P. and RÉNYI, A. (1966), On the Existence of a Factor of Degree one of Connected Random Graphs, *Acta Math. Acad. Sci. Hung.* **17**, 359–379.

ERDŐS, P. and SPENCER, J. (1972), Imbalances in *k*-Colorations, *Networks.* **1**, 379–385.

ERDŐS, P. and SZEKERES, G. (1935), A Combinatorial Problem in Geometry, *Compositio Math.* **2**, 463–470.

FELLER, W. (1957), *An Introduction to Probability Theory and its Applications*, Vol. I., John Wiley and Sons, New York.

GALLAI, T. (1963), Kritische Graphen I. *Magyar Tud. Akad. Mat. Kutató Int. Közl.* **8**, 165–192 (esp. 172–175).

GOODMAN, A. W. (1959), On Sets of Acquaintances and Strangers at any Party, *Amer. Math. Monthly* **66**, 778–783.

GORDON, Y. and WITSENHAUSEN, H. S. (1972), On Extension of the Gale–Berlekamp Switching Problem and Constance of l_p-spaces, *Israel J. of Math.* **11**, 211–229.

GRAVER, J. E. and YACKEL, J. (1968), Some Graph Theoretic Results Associated with Ramsey's Theorem, *Journal of Combinatorial Theory* **4**, 125–175.

GUY, R. and ZNÁM, S. (1969), A Problem of Zarankiewicz, in *Recent Progress in Combinatorics*, Ed. by TUTTE, W., Academic Press, New York—London.

HALL, M. (1967), *Combinatorial Theory*, Ginn and Blaisdell, Walthem-Toronto-London.

HAJNAL, A. (see. P. Erdős).

HANANI, H. (see P. Erdős).

HANANI, H. (1960), On Some Quadruple Systems, *Canad. J. of Math.* **12**, 145–157.

HANANI, H. (1961), The Existence and Construction of Balanced Incomplete Block Designs, *Annals of Math. Statistics* **6**, 362–386.

HARARY, F. (see L. W. Beineke).

HARNER C. (see R. C. Entringer).

HERZOG, M. and SCHOHNHEIM J. (1972), The B_r Property and Chromatic Numbers of Generalized Graphs, *J. Comb. Theory* (B) **12**, 41–49.

KATONA, G., NEMETZ, T. and SIMONOVITS, M. (1964), On a Graph-problem of Turán (in Hungarian), *Mat. Lapok* **15**, 228–238.

KELLY, J. B. and KELLY, L. B. (1954), Paths and Circuits in Critical Graphs, *American J. of Math.* **76**, 786–792.

KELLY, L. B. (see Kelly, J. B.).

KOMLÓS, J. and SULYOK, M. (1969), On the Sum of Elements of ± 1 Matrices, in the Proceedings of the Combinatorial Colloquium in Balatonfüred, Hungary, 1969.

KÖNIG, D. (1950), *Theorie der endlichen und unendlichen Graphen*, Chelsea, New York.

KORŠUNOV, A. D. (1971), The Number of the Nonisomorphic Subgraphs in an n-vertex Graph (in Russian), *Mat. Zametki* **9,** 263, 272; English translation, *Math. Notes* **9,** (1971) 155–160.
KŐVÁRI, T., SÓS, T. V. and TURÁN. P. (1969), On a Problem of K. Zarankiewicz, *Colloqium Mathematicum* **3,** 50–57.
LINDSTRÖM, B. (1965), On a Combinatorial Problem in Number Theory, *Can. Math. Bull.* **8,** 477–490.
LOVÁSZ, L. (1968), On Chromatic Number of Finite Set-Systems, *Acta Math. Acad. Scient. Hung.* **19,** (1–2), 59–67.
MILLER, E. W. (1937), On a Property of Families of Sets, *Comptes Rendus Varsovic* **30,** 31–38.
MILLS, W. H. (see D. G. Cantor).
MOON, J. W. (see P. Erdős).
MOON, J. W. (1968), *Topics on Tournaments,* Holt, Rinehart, and Winston, New York.
MOSER, L. (see P. Erdős).
MOSER, L. (1960), On a Theorem of van der Waerden. *Canadian Mathematical Bulletin* **3,** 23–25.
MOTZKIN, T. S. and STRAUSS, E. G. (1965), Maxima for Graphs and a New Proof of a Theorem of Turán, *Canad. J. Math.* **17,** 533–540.
NEMETZ, T. (see G. Katona).
OTTER, R. (1949), The Multiplicative Process, *Annals of Math. Stat.* **50,** 206–224.
RADO, R. (see P. Erdős).
RAMSEY, F. P. (1930), On a Problem of Formal Logic, *Proc. of the London Math. Soc.*, 2nd series **30,** 264–286.
RÉNYI, A. (see P. Erdős).
RIEMAN, I. (1958), Über ein Problem von K. Zarankiewicz, *Acta Math. Acad. Sci. Hung.* **9,** 269–279.
ROTH, K. F. (1964), Remark Concerning Integer Sequences, *Acta Arithmetica* **9,** 257–260.
RYSER, H. J. (see R. H. Bruck).
RYSER, H. J. (1963), *Combinatorial Mathematics,* Carus Monograph Number 14, John Wiley and Sons, New York.
SCHMIDT, W. M. (1962), Two Combinatorial Theorems on Arithmetic Progressions, *Duke Math. J.* **29,** 129–140.
SCHMIDT, W. M. (1964), Ein kombinatorisches Problem von P. Erdős und A. Hajnal, *Acta Math. Acad. Sci. Hung.* **15,** 373–374.
SCHÖNHEIM, J. (see M. Herzog).
SCHÖNHEIM, J. (1966), On Maximal Systems of r-Tuples, *Studia Sci. Math. Hung.* **1,** 363–368.
SIMONOVITS, M. (see G. Katona).
SPENCER, J. (see T. A. Brown and P. Erdős).
SPENCER, J. (1971*a*), Optimal Ranking of Tournaments, *Networks* **1,** 135–138.
SPENCER, J. (1971*b*), A Remark on Coloring Integers, *Canad. Math. Bull.* **14,** 45–47.
SPENCER, J. (1972*a*), Turán's Theorem for k-graphs, *Discrete Math.* **2,** 183–186.
SPENCER, J. (1972*b*), Random Regular Tournaments, *Rand Corp. Paper,* P-4479-1.
STORER, T. (1967), *Cyclotomy and Difference Sets,* Markham.
STRAUSS, E. G. (see T. S. Motzkin)
SULYOK, M. (see J. Komlós).
SZEKERES, E. and SZEKERES, G. (1965), On a Problem of Schütte and Erdős, *Math. Gazette* **49,** 290–293.
SZELE, T. (1943), Kombinatorikai vizsgálatok az irányított teljes gráffal kapcsolatban, *Mat. Fiz. Lapok* **50,** 223–256. For a German translation, see Kombinatorische Untersuchungen über gerichtete vollständige Graphen, *Publ. Math. Debrecen* **13,** 145–168 (1966).
TARRY, G. (1900), Les problèmes des 36 officiers, *C. R. Assoc. Franç. Av. Sci.* **1,** 122–123 and **2,** 170–203 (1901).

Turán, P. (1941), Egy gráfelméleti szélsőérték feladatról, *Mat. Fiz. Lapok* **48,** 436–452; see also On the Theory of Graphs, *Colloq. Math.* **3,** 19–30 (1954) and Motzkin, Strauss (1965).

Turán, P. (1961), Research Problems, *Publ. Math. Inst. Hung. Acad. Sci.* **6,** 417–423.

Tutte, W. T. (1947), (under pseudonym Blanche Descartes), A Three-colour problem, *Eureka* (April 1947, solution March 1948) and Solution to Advanced Problem No. 4526, *Amer. Math. Monthly*, **61,** 352 (1954).

Tutte, W. T. (1947), The Factorization of Linear Graphs, *J. London Math. Soc.* **22,** 107–111.

van der Waerden, B. L. (1952), Beweis einer Baudet'schen Vermutung, *Niew Archief voor Wiskunde*, **15,** 212–216.

Vizing, V. G. (1968), Some Unsolved Problems in Graph Theory (in Russian), *Uspekhi Matem. Nauk.* **23,** 117–134.

Witsenhausen, H. S. (see Y. Gordon)

Yackel, J. (see J. E. Graver).

Yackel, J. (1972), Inequalities and Asymptotic Bounds for Ramsey Numbers. *J. Comb. Theory* **13,** 56–68.

Znám, S. (see R. Guy).

Zykov, A. A. (1949), On some Properties of Linear Complexes (in Russian), *Russian Math. Sbornik, N. S.* **24** (66), 163–188.